THE SCIENCE OF
STRONG WOMEN

THE SCIENCE OF
STRONG WOMEN

THE TRUE STORIES BEHIND YOUR FAVORITE FICTIONAL FEMINISTS

RHIANNON LEE

ILLUSTRATED BY ALICE NEEDHAM

Skyhorse Publishing

THIS BOOK IS INTENDED TO BE INCLUSIVE TO ANYONE WHO IDENTIFIES AS A WOMAN—BECAUSE THE MORE STRONG WOMEN THERE ARE ON THIS PLANET THE BETTER!

Skyhorse Publishing books may be purchased in bulk at special discounts for sales promotion, corporate gifts, fund-raising, or educational purposes. Special editions can also be created to specifications. For details, contact the Special Sales Department, Skyhorse Publishing, 307 West 36th Street, 11th Floor, New York, NY 10018 or info@skyhorsepublishing.com.

Skyhorse® and Skyhorse Publishing® are registered trademarks of Skyhorse Publishing, Inc.®, a Delaware corporation.

Visit our website at www.skyhorsepublishing.com.

10 9 8 7 6 5 4 3 2 1

Library of Congress Cataloging-in-Publication Data is available on file.

Cover design by David Ter-Avanesyan
Cover illustrations by Alice Needham
Interior design and layout by Chris Schultz

Print ISBN: 978-1-5107-7087-4
Ebook ISBN: 978-1-5107-7088-1

Printed in the United States of America

Here's to strong women.
May we know them,
may we love them,
may we be them,
may we celebrate them.

Foreword

Throughout these pages you will find the fictional characters I treasure the most, all of whom have helped shape me in a small way, expanded my world, and taught me what it means to be a strong woman.

While writing this book, I found it extremely hard to narrow the scope down to just fifty inspiring female characters. A glance at my overflowing bookcase or a look back in my Netflix history shows just how obsessive my consumption of strong fictional women is! I wanted to feature characters that not only represent a variety of genres, but also a variety of backgrounds and periods in history, with each character representing different qualities and perspectives on what it means to be a strong woman. It would be easy, for example, to fill these pages with feminist superheroes, but there is a lot more to being a strong woman than donning a flattering spandex suit and high kicking your problems away.

The time in history and the places in which these characters exist are key to understanding what make them strong women. With our views on feminism and gender constantly evolving, what made sense two hundred years ago can seem wildly outdated by today's standards. For example, you won't see Elizabeth Bennet burning her corset or telling Mr. Darcy to shove his outdated gender views up his ass. Nonetheless, for her time and place she was a truly radical feminist icon.

In truth, there is no perfect fictional feminist out there, no goddess-like woman who exhibits every desired trait of femininity and strength (and if there was, what a boring character she would be anyway). The characters featured here are flawed individuals, all figuring it out as they go along—much like we are.

Contents

Introduction

I expect that everybody has read a book or seen a show at some point and really identified or felt drawn to a certain character. That feeling of being immersed and seeing the world through a different perspective, gaining an insight into what makes that character tick, is a luxury we seldom have in our daily lives. Even if you are surrounded by friends and family members who are emotionally honest, it is human nature to conceal some of our innermost feelings. Living in a world where everyone says they're "fine," fictional characters can be some of the most transparent individuals in our lives, making them easier to relate to.

The fictional stories we grew up with can play a huge role in shaping our lives and leave a lasting impression. Whether it's in books or on-screen, fiction enables us to experience new worlds and characters we wouldn't normally meet and see the inner workings of other people's minds—all from the comfort of our own sofa. The fictional environment that we are exposed to as we grow helps determine our values, ambitions, and imagination so it makes sense that for many young people, the female characters they encounter influence their own views on femininity, feminism, and what it means to be a strong woman.

In this book I will introduce you to the stories that influenced me throughout my life and to the exceptional fictional women I have met along the way. Whether it is having a crush on someone at school, overcoming adversity, dealing with racism/sexism, or surviving a dystopian future (thanks, global pandemic!), there is a fictional character for every experience. And by reading about or watching their triumphs, fears, and struggles, it can help you with your own.

A strong woman is not just a kick-ass lady who solves her own problems and goes off on her own adventures. She also encapsulates many qualities such as inner strength, bravery, intelligence, determination, and resilience. In a world where much of how a society perceives its women is shaped by the media, it's important for books, TV shows, and films to offer positive role models for girls and women to look up to, who can hold their own, and are not just waiting for their knight in shining armor to deliver them to their happily ever after! There are so many strong women who meet these criteria, and the following pages include just a short selection of fifty.

NOT ALL SUGAR AND SPICE AND EVERYTHING NICE . . .

Particularly in children's literature and media, it is hard to escape the overplayed gender stereotypes of princesses dressed in pink waiting for their perfect Prince Charming to come along and save the day. These lazy damsel in distress narratives have been a blueprint for much of what many of us consumed as a child. From *Cinderella* and *Sleeping Beauty* to *Snow White*, the building blocks for what was meant to interest young girls were always the same: A demure and delicate woman valued only for her exceptional beauty who has no control over her own destiny and needs saving from her hardships by a man. This played out storyline had remained popular since Victorian times when most stories aimed at young girls were supposed to instill a strong moral code with tales of wicked women being punished for their unruly willfulness while virtuous women are rewarded with a happily ever after. I'm happy to report that these storylines are destined for the history books, with even Disney undergoing a feminist reboot in recent years with films like *Frozen*, *Moana*, and *Encanto* showing the next generation that girls are perfectly capable of solving their own problems.

Many of the characters featured in this book come from books which have, over the years, been repeatedly banned from school classrooms or entire countries. For example, *The Color Purple*, *The Scarlet Letter*, and *To Kill a Mockingbird*, to name a few. I find this extremely telling of society's fear of narratives that feature strong women and the often "unpalatable" stories they have to tell. By choosing to exclude a narrative,

it sends a clear message that society is actively trying to silence women. Unfortunately, the problem still persists, and as recently as 2020 a high school in Georgia pushed for the removal of Margaret Atwood's feminist classic *The Handmaid's Tale* from the curriculum due to vulgarity and sexual overtones.

If we censor the stories that don't have a happily ever after, we are not getting a true snapshot of the human experience. Life can be devastating, cruel, and painful, something most women are very aware of. But if these experiences are seldom reflected, how are we as women expected to feel represented?

FEMINISM IN FICTION

Throughout the course of history, literature and film are embellished with spectacular works of fiction, theory, and criticism all revolving around one thing: feminism. Even when women had less of a voice in the literary world, you still see strong female protagonists.

The history of women in fiction is a complex one. Historically, women were the storytellers, the ones who would enchant children with tall tales to impart wisdom and life lessons to the next generation. For thousands of years, mothers, aunts, and grandmothers held the narrative, telling stories by the fire or in the nursery. The problem was that for the better part of history, men were the only ones allowed to write down and record these narratives. Until the mid-nineteenth century, women were discouraged from education and forbidden to publish or make a career in storytelling. It all boils down to the point Virginia Woolf made in her essay *A Room of One's Own*, that society has different expectations for women and as a result creates no space for them. Woolf argued this point with Judith, the fictional sister of Shakespeare. Judith may have had the same aptitude for playwriting as her brother William, maybe even greater. However, as Judith was banned from receiving a formal education and unable to seek her own fortune, the world never knew her or her stories.

The Brothers Grimm, often labeled as the fathers of fairy tales, published stories that were not plucked from their own imaginations, but were in fact a collection of folk tales they composed from the words of generations of mothers and grandmothers. It can be argued that the

Brothers helped save these stories from fading into the passages of time, preserving them for many generations to enjoy. However, it is undeniable that the Brothers profited hugely from their ability to publish and be heard, something that would not have been possible for two sisters to do at the time.

Enter the world of film and television where women, especially women of color, have been underrepresented. According to a study conducted in 2014 by the Geena Davis Institute on Gender in Media[*] which analyzed 120 films made worldwide from 2010 to 2013, only 31 percent of named characters were female and 23 percent of the films had a female protagonist or co-protagonist while only 7 percent of directors were women. This study is not alone in reflecting the underrepresentation of women in film. Another study published by the Media, Diversity & Social Change Initiative[†] examined the 700 top-grossing films from 2007 to 2014 and found that only 30 percent of the speaking characters were female, which translates to a gender ratio of 2.3 men to every woman.

In 1985, comic strip illustrator Alison Bechdel proposed a way of assessing the gender equality of a film. She outlined three criteria to gauge whether women actively feature in a film, which are:

1. The presence of two named female characters.
2. These two female characters have to talk about something . . .
3. . . . that has nothing to do with a man.

This test isn't used to measure feminist content, just the mere existence of women in a movie. For example, you could have a film which features, for just thirty seconds, two women named Becky and Jennifer, having a conversation about how overrated they feel feminism is, pining for the good-old-days before women were weighted down with the pressures of voting or autonomy over their own bodies. This one scene would cause the film to pass the Bechdel test. Sounds simple, right? Wrong. In a 2021

---

[*]   Smith, S. *et. al.*: Gender Bias without Borders: An Investigation of Female Characters in Popular Films across 11 Countries. Geena Davis Institute on Gender in Media. 2015.

[†]   Smith, S. *et al.*: Inequality in 700 Popular Films: Examining Portrayals of Gender, Race & LGBT Status from 2007 to 2014. Media, Diversity & Social Change Initiative. 2015.

article from the Journal of the Data Visualization Society[*], it was reported that 40 percent of movies still fail the Bechdel test, and among those, 10 percent fail to satisfy even one of the three criteria.

BRIEF HERSTORY OF FEMINISM

Feminism is often described as happening in waves and, just like coastal erosion, each new wave hopes to take a chunk out of the gargantuan sheer cliff which is the patriarchy. Below is a brief summary of the history of feminism.

THE FIRST WAVE

Historically, the mid-seventeenth century is often cited as the beginning of the first wave of feminism, as it features some of the first texts that discuss the women's liberation movement. However, this feels slightly unfair to most of human history, as scientists have recently revealed through the analysis of hand size that three quarters of cave paintings were created by women[†]. These cave paintings often depict game animals such as bison, reindeer, horses, and woolly mammoths. It had always been assumed that these images were made by male hunters, as a way to chronicle their skills, but perhaps this was the first wave of feminism, with cave women celebrating their own hunting skills or being the bearers of knowledge. These simplistic drawings are some of the earliest evidence of storytelling. With little evidence to go on, we might never know why women were the predominant artists in the Paleolithic Era, but the cave paintings made 2 million years ago give a faint echo to the voices of these early women.

A voice which was very clear in its disdain for how women were portrayed was that of Mary Wollstonecraft (1759–1797), an English writer and philosopher. In her 1792 essay *A Vindication of the Rights of Women*, she wrote "I may be accused of arrogance; still I must declare, what I firmly believe, that all the writers who have written on the subject of female education and manners . . . have contributed to render women more

---

[*] Viswanathan, G. What Is the Bechdel Test and What Is Its Relevance to Today's Film Industry? Journal of the Data Visualization Society. July 2021.

[†] Snow, D. Sexual Dimorphism in European Upper Paleolithic Cave Art. *American Antiquity*, 2013;78(4), 746-761.

artificial, weaker characters, than they would otherwise have been; and, consequently, more useless members of society." Wollstonecraft's essay brought the discussion of a woman's place in society into the mainstream.

The term feminism itself was first coined in 1837 by French philosopher Charles Fourier. It originally referred to feminine qualities or character, but toward the end of the nineteenth century it became inextricably linked with equal rights for women and the suffrage movement.

The first wave really started at New York's Seneca Falls Convention of 1848, the first women's rights convention. This two-day convention produced the Declaration of Sentiments (based on the Declaration of Independence) and was primarily authored by Elizabeth Stanton. In it she wrote "The history of mankind is a history of repeated injuries and usurpation on the part of man toward woman, having in direct object the establishment of an absolute tyranny over her." This declaration called for women to recognize the injustice they were held under, and it was this declaration that set the fire under the heavily corseted dresses of many women to start the suffrage movement. However, votes for women took more than seventy years, with Congress only passing it in 1920, sadly something Elizabeth Stanton never lived to see.

While votes for women may have been the beginning for the suffragettes, other minorities were also demanding equal freedoms such as the end of enslavement. Perhaps the most notable woman to have been active in both movements was abolitionist Sojourner Truth, a former slave who was the first black woman to successfully sue a white slave owner to reclaim her children. Truth's famous speech "Ain't I a Woman?" was in response to the all-male jury who presided over her trial that discussed the nature of women as being delicate and should be treated as such. Truth's legacy is linked to the foundations of the theory of feminism and many years later her speech was performed again by actress Kerry Washington as part of Voices of People's History of the United States.

The first wave of feminism influenced the popular culture of its time. Mary Shelley, English Gothic novelist and daughter of feminist icon Mary Wollstonecraft, published her novel *Frankenstein* in 1818, first anonymously but then under her own name in 1823. *Frankenstein* contained many references to her mother's ideology and work. It is also clear to see the influence of first-wave feminism in the works of Jane Austen, who wrote female-centered stories often portraying the struggles of a woman's place in society. See page 89 for a full discussion of the feminist icon that is Elizabeth Bennet and why she is still such a beloved fictional heroine to this day.

THE SECOND WAVE

Feminism's second wave (often referred to as "women's lib") is character-ized by the women's rights movement of the sixties and seventies. This wave focused on women pursuing a career, reproductive rights, equal pay, and highlighting the horrific violence women were experiencing. Spurred on by activist Betty Friedan's 1963 book *The Feminine Mystique*, many women responded to the "problem that has no name." Friedan was dismissed from her journalism job in 1952 because she became pregnant with her second child. Forced to be a stay-at-home mom, Friedan began to question why women were expected to face the brunt of childcare, putting their own education and careers to the side in order to raise a family. This opinion was echoed the more Friedan talked to other women, yet she was floored that she had never seen it reported before. Instead, media and society were pushing the message that a woman can only be fully content in life if she is raising children.

The second wave came to its peak with French philosopher Simone de Beauvoir's groundbreaking book *The Second Sex*. Simone declared that women in society are labeled as "Other." She writes, "The category of the *Other* is as primordial as consciousness itself. In the most primitive societies, in the most ancient mythologies, one finds the expression of a duality—that of the Self and the Other." Her works talk about women's position in society as secondary, almost as an afterthought, and how this has become the norm through generations. The messages of Friedan and de Beauvoir, combined with the changing political climate of the 1960s, created a generation of women who were willing to expand upon the work from the previous generation of women.

The second wave of feminism saw some historic events emerge, such as the mass production of contraceptives, laws in place to protect the equal rights of women, and more women being involved in politics. These events had an influence on the pop culture of the time with television shows such as *The Lucy Show* and *Bewitched* showing women as independent individuals.

However, criticism of the second wave of feminism was that race, sexuality, and social class were not equally addressed. Patricia Hill Collins's book *Black Feminist Thought* does not specifically call out the lack of inclusion in the second wave movement, but it does address the need to place the voice of women of color at the forefront of feminist theory, as often they are the ones who face the greatest prejudice. This prejudice is still felt today with one 2020 study on the earning differences by gender, race, and ethnicity showing that for every dollar a white man earns, a white woman earns $0.79 cents, while a black women earns 20 percent less than the white woman (altogether 37 percent less than the white man) at $0.63 cents.

The works of strong women such as Patricia Hill Collins, Gloria Anzaldua, and Audre Lorde contributed to what is known as the third wave of feminism.

THE THIRD WAVE

Although it is unclear where the second wave ended and the third wave began, bestselling book *Manifesta: Young Women, Feminism, and the Future* is accredited with bringing the third wave into the mainstream. Feminism's influence over pop culture was pivotal during this time. A good example is in the music scene of the mid 1990s with the "Riot Grrrl" movement. Starting as an underground feminist punk movement, it quickly spread to over twenty-six countries. This movement allowed a new generation to become politically conscious and fight out against the patriarchy. However, unlike the women of the second wave who grew up in the sixties, the women of the third wave are said to have been raised in a "post-feminism era," not having to fight for the rights to play sports, get birth control, or pursue a career. The third wave had the biggest impact on Hollywood when the moviemaking industry realized there was a lot of money to be made in feeding the high demand for female-led shows. This has resulted in some of our most beloved fictional feminists like Princess Leia and Buffy Summers.

The third wave also shifted its focus to non-Western women in the discussion of women's rights.

THE FOURTH WAVE

The fourth wave brings us into the twenty-first century, and although the attention may still be on questions raised by earlier waves, through the use of social media, feminism found a new platform. The hashtags #MeToo and #TimesUp rocked the world in 2006, spotlighting the sexual harassment and abuse women were subjected to on a daily basis. The media covered some of the biggest sex scandals in Hollywood, although the effects were felt in many countries, with politicians taking notice of this global movement.

In 2012, British writer and activist Laura Bates started up the Everyday Sexism project (www.everydaysexism.com) to record women's impressions of the ingrained sexism they experience. The project saw over a hundred thousand women's experiences of sexism and discrimination being recorded, from the classroom to the workplace, in the media and in public spaces.

Fourth wave feminism also reevaluated the definition of who a woman is and has become more LGBTQ+ inclusive, acknowledging the previously marginalized trans community.

THE SCIENCE OF
STRONG WOMEN

PART ONE

Girls Who Run the World

There's no power like girl power! And that's certainly true for what we read and watch during our formative years. The characters we encounter and relate to can have a real lasting effect on us. They influence how we develop ideas, define what motivates us, and broaden our sense of self and others.

From *The Very Hungry Caterpillar* and *The Gruffalo* to *Peter Rabbit* and *Babar*, children are exposed from the youngest age to a clear gender disparity. A comprehensive study[*] in 2011 on popular children's literature over the past century found that 57 percent of published children's books feature a male protagonist, compared to just 31 percent which feature a female protagonist. Even when a girl does manage to score a lead role in a story, they are often very much a stereotype (i.e., a damsel in distress or a one-dimensional side-kick), to be a sounding board for the other male characters who are having the adventure. The lack of strong female characters in fiction not only sends outdated messages to young girls, but also to the young boys who read these same stories. We need to help boys see that it's okay for a girl to be stronger and more powerful than them, and that it doesn't make them weak in comparison.

---

[*] McCabe J., *et al.*: Gender in Twentieth-Century Children's Books: Patterns of Disparity in Titles and Central Characters. *Gender & Society*. 2011;25(2):197-226

In 2018, the *Guardian* conducted an in-depth study of Nielsen BookScan's one hundred top-selling children's picture books. Their study revealed that across these titles, there were three male characters for every two females, while central characters with speaking parts were 65 percent male and only 35 percent female.

Even when books do feature women and girls, many of the most popular stories we remember fondly from our early years, in retrospect, can have misogynistic undertones. Take the popular Mr. Men series of picture books originally written by Roger Hargreaves and first published in 1971. It was not until 1980 that Little Miss characters were added to the mix, and even then, these female characters were often rife with gender stereotypes. While the Mr. Men characters got to be smart, funny, and brave, the Little Miss characters were portrayed as naughty, scatterbrained, and vain, with characters such as Little Miss Princess, Little Miss Tidy, and Little Miss Bossy. Today, the Mr. Men universe has seen the welcome addition of new characters like Little Miss Inventor to interest more girls in science and engineering.

Gender imbalance and the reinforcing of gender stereotypes in children's books is one of the earliest stumbling blocks on the road to equality. Therefore, Part One of this book celebrates the strong female protagonists in children's books and media and the lessons they impart on us.

ALICE, ALICE'S ADVENTURES IN WONDERLAND SERIES

"Alice had begun to think that very few things indeed were really impossible."

LESSON LEARNED

The imagination is a powerful thing, so never stop wondering.

PLOT

Alice's Adventures in Wonderland starts with Alice observing a white rabbit scurrying down a rabbit hole. Her curiosity peaked, she decides to follow. She arrives in Wonderland where she encounters many strange characters, including the Cheshire cat, who advises her to attend a tea party thrown by the Mad Hatter. After a bizarre experience where the Mad Hatter tries to cut her hair, Alice runs away and finds herself in a formal garden surrounded by servants painting roses red to appease the foul-tempered Queen of Hearts. Alice then finds herself at the mercy of the Queen's court, accused of a crime she did not commit. When the Queen orders her execution, Alice awakens just in time to realize it was all just a bizarrely vivid dream.

FEMINIST ICON

Alice's Adventures in Wonderland was written by British author (and mathematician, oddly enough) Lewis Carroll during the Victorian era. In Victorian times, gender roles were pretty inflexible and women were supposed to be meek, compliant, and subservient to men. However, Lewis Carroll made the protagonist in this series a rambunctious seven-year-old girl who did not conform to these notions about how little girls were supposed to behave. She was a curious adventurer who faced her fears and used her brain to overcome obscure obstacles. Sure, the whole thing may have been just a dream, but conquering the fears that live in your subconscious is no easy feat and something we could all benefit from!

Who doesn't fall down a rabbit hole and enter into an illogical world from time to time? For me, Alice has always been a great role model. Her imagination allows her to adapt to continually changing circumstances, and her wits and determination help her counteract the crazed characters she encounters. Alice may only be a child, but by conquering Wonderland she shows us that the patriarchy doesn't stand a chance!

ABOUT THE AUTHOR

Lewis Carroll is actually a pen name for Charles Dodgson. Charles attended Oxford University where he studied mathematics and went on to publish eleven purely mathematical books. He first pitched the idea of *Alice's Adventures in Wonderland* to the three children of his boss, Henry Liddell, the dean of Christ Church Oxford. It was one of Henry's children, Alice, who inspired the name of our heroine.

FACTS

- After reading and loving *Alice's Adventures in Wonderland*, Queen Victoria suggested that Carroll dedicate his next book to her. His next work, *An Elementary Treatise on Determinants: With Their Application to Simultaneous Linear Equations and Algebraic Equations*, was dedicated to the monarch—perhaps not the page turner she was expecting!
- The series was banned in China in 1931 on the grounds that "animals should not use the human language."
- Since the series was published in 1865, it has been translated into 176 languages and has never been out of print.
- Walt Disney based the 1951 all-animated musical film *Alice in Wonderland* on *Alice's Adventures in Wonderland*, but he had originally planned for an animated mixed with live-action film.

MARY LENNOX,
THE SECRET GARDEN

"*She made herself stronger by fighting with the wind.*"

LESSON LEARNED

When things are neglected, they wither and die, but when they are cared for, they thrive.

PLOT

The story starts with Mary Lennox in India during a cholera outbreak that wipes out her parents and their servants. During this crisis, Mary is forgotten. She is later found in her nursery and shipped off to her uncle's sprawling mansion where she is essentially on her own and has to make the best of a bad situation. During her days of wandering the vast estate, Mary discovers a locked, walled-off garden with a hidden past. As any stubborn, ignored child of the early 1900s would do, Mary vows to restore the garden to its former glory, despite the potential consequences. Aided by grounds staff and her frail cousin, she learns the value of hard work, grit, determination, and friendship.

FEMINIST ICON

The secret garden was first published in 1911, a time when society was entering a period of considerable industrial change and, as a consequence, social upheaval. This led to an enduring tale of a girl growing up, overcoming obstacles, and embracing her inner eco-warrior. *The Secret Garden* brings us the story of Mary Lennox, an orphaned protagonist with a twist. Unlike the good-hearted, put-upon creatures from the likes of *Oliver Twist* or *Cinderella*, Mary is spoiled, rude, and sometimes violent.

It is rare in children's literature to have a disagreeable and self-centered lead protagonist, but that is what makes Mary all the more interesting! Her wild spirit is somewhat lessened by the end of the story; some argue to conform to Victorian standards of the time. However, I argue that making friends and being at one with nature will mellow even the most irritable. Classic children's literature is enriched by Mary Lennox, who represents the disagreeable cynic in many of us.

ABOUT THE AUTHOR

Frances Hodgson Burnett was a British American novelist born in 1849. Originally born in Manchester, England, Frances's family moved across the Atlantic to Tennessee after her father's death when she was three. Burnett traveled for much of her life and married twice. After divorcing her first husband, she married an aspiring actor ten years her junior in Italy. However, the press had a field day with the age difference and regarded Frances's second marriage as the greatest mistake of her life. Thankfully, we live in much more enlightened times now when a successful woman marrying a much younger man is not so scandalous.

FACTS

- The setting of *The Secret Garden* was inspired by a real garden at Great Maytham Hall in Kent, England, which Burnett rented in 1898. Picture a Downton Abbey-esque manor with a beautiful walled garden.
- In 1936, a memorial sculpture was erected in Burnett's honor in Central Park's Conservatory Garden depicting the characters Mary and Dickon from *The Secret Garden*.
- *The Secret Garden* was originally published in a magazine for adults before being published as a book and rebranded as a story for children.
- At the time of Burnett's death, *The Secret Garden* was one of her least popular stories. It didn't compare in popularity to other works by Burnett such as *Little Lord Fauntleroy* or *The Little Princess*.

LUCY PEVENSIE,
THE CHRONICLES OF NARNIA SERIES

"I think—I don't know—but I think I could be brave enough."

LESSON LEARNED
A girl can save the world (with a little help from a magical lion).

PLOT
During the World War II bombing of London, four siblings, Lucy, Peter, Susan, and Edmund, are evacuated to safety in a country house. One day the youngest of the siblings, Lucy, finds a wardrobe which transports her to a magical world called Narnia. Upon convincing her siblings of the existence of Narnia, they all travel there together, meeting many magical creatures along the way. However, everything is not all rosy in Narnia and the siblings, with the help of a magical lion, must perform some improbable feats to rescue the kingdom from sinister forces.

FEMINIST ICON
C. S. Lewis would roll in his grave if he knew he was being attributed as the creator of one of the most iconic feminist characters in children's literature. As a very conservative Christian, C. S. Lewis held pretty bigoted views about women and feminism, made clear through his other works. It pains me to say that even my favorite Narnia book growing up, *The Lion, the Witch, and the Wardrobe*, includes some frank sexism, with Pevensie boys Peter and Edmund getting to fight in battles while girls Susan and Lucy merely watched from the sidelines. Despite the fact that Susan is a skilled archer, Aslan the Lion (and thinly veiled Christ-figure and voice of God himself) comments "battles are ugly when women fight." As if they're somehow pretty when men do? The logic is not as strong as you'd expect from the deity, perhaps, but the point is clear enough: no women allowed in the military in Narnia.

However, putting aside the sexism for one second, there are some feminist gems within The Chronicles of Narnia series which I cling to. Lucy is the real protagonist of the book. She was the first to find Narnia by going through the wardrobe, encountering disbelief from her siblings on returning; they think she's telling tales, or that she's insane. However, Lucy proves she is not just a silly little girl and goes on to lead the others to Narnia. Lucy possesses a spirit of adventure and is extremely open-minded and caring for all the creatures who live in Narnia. She can also be seen as a symbol of courage and perseverance.

ABOUT THE AUTHOR

C. S. Lewis was a British writer born in 1898 in Belfast, Ireland. He attended the University of Oxford where he met a number of famous playwrights and poets. His journey into academia was suspended, however, due to the outbreak of WWI, which led to Lewis joining the British army and fighting in France. After being injured and sent back to England, Lewis finally graduated from university and started his career in writing. In total, Lewis wrote more than thirty novels in his lifetime. To date, The Chronicles of Narnia series (his most popular work) has sold over one hundred million copies and has been transformed into major motion pictures.

FACTS

- Lewis and J. R. R. Tolkien were in a writing group called The Inklings. The strictly men's-only group would meet every Monday morning to talk about writing.
- Even with the help of his prodigious writing group, it took Lewis ten years to finish *The Lion, the Witch, and the Wardrobe* (something that the procrastinator in all of us can relate to).
- Lewis actually destroyed the first version of *The Lion, the Witch, and the Wardrobe*, as his friends' reaction to the story was discouraging, to say the least. Lewis said in a letter, "It was, by the unanimous verdict of my friends, so bad that I destroyed it."

GEORGE (GEORGINA), THE FAMOUS FIVE SERIES

"She's wonderful. She's the bravest girl I ever knew."

LESSON LEARNED

There is freedom in defying the "feminine" stereotypes and embracing our inner tomboy.

PLOT

The Famous Five was a series of adventure books surrounding a group of young children: Julian, Dick, Anne, Georgina (George), and their dog Timmy. The stories take place during summer vacation when the children all return from their respective boarding schools. Every time they meet, they get caught up in an exciting adventure, often involving criminals or lost treasure.

FEMINIST ICON

In 2008, Enid Blyton, author of the Famous Five series and many other children's books, was voted Britain's best loved author. Children all around the world grew up reading her stories. In her Famous Five series, one character in particular stood out to me: Georgina (or George, as she demands to be called). George was hell-bent on finding adventure and challenged traditional gender roles. Not only did she have a wonderful dog, Timmy (who often saved the day), but she also had short hair, climbed ropes, wore shorts, swam, sailed, rowed boats, and owned an island! Everything about her appealed to my inner tomboy.

George taught me that girls could do anything that boys could, and showed me the fun that could be had camping and exploring the outdoors. It was George who inspired me to join the Boy Scout troop in my village rather than joining the Girl Scouts. For me it was a no-brainer, the Boy Scouts would get to sleep outdoors, play with axes, and build campfires, whereas the Girl Scouts would have sleepovers in the village hall and make crafts, with not a lethal weapon to be found! I knew exactly which one George would join, and so with that in mind, my best friend and I set out to become the first girls to join our village Boy Scout troop, thanks to George. I still have the axe scars to prove it!

ABOUT THE AUTHOR

Enid Blyton was born in England in 1897. She wrote books on a wide range of topics but is best remembered for her Famous Five series. However, a deeper look at her books brings up elitist, racist, sexist, and homophobic language and themes. The argument has been made recently in the press and social media that she was writing in a style that was "of her time." However, at the height of her fame in the 1960s, Blyton's work started to be rejected from major publishers on the grounds that it was xenophobic in nature.

FACTS

- Enid Blyton's books have sold more than six hundred million copies.
- She was described as a very difficult woman, with her own daughter, Imogen Blyton, describing her as "arrogant, insecure, and pretentious."
- In recent years, the Famous Five series has resurfaced with Ladybird Books rewriting the series for adults. Titles now include *Five on Brexit Island, Five Go Gluten Free,* and my own personal favorite, *Five Give Up the Booze.*
- Enid Blyton was said to be partial to a bit of naked tennis in her free time.

PIPPI LONGSTOCKING, PIPPI LONGSTOCKING SERIES

Astrid Lindgren: *"He's the strongest man in the world."*
Pippi Longstocking: *"Man, yes," said Pippi, "but I am the strongest girl in the world, remember that."*

LESSON LEARNED
Don't live by anyone's rules but your own.

PLOT
Pippi Longstocking is a red-haired, pigtailed, freckled, unconventional, and superhumanly strong (she can lift her horse one-handed) nine-year-old girl. Her mother died when she was a baby and her father, a sea captain, has seemingly vanished at sea. However, this does not phase Pippi who lives alone in a big house with her pet monkey, horse, and a suitcase filled with pieces of gold. Along with her two best friends, Annika and Tommy, she gets up to all sorts of mischief and never compromises on staying true to herself.

FEMINIST ICON
Every young child should be introduced to the original rebel girl with a hunger for freedom: Pippi Longstocking. Pippi does not just have fun adventures; she also embodies what every young child (and many adults!) dreams of: superhuman strength, independence, and a pet monkey! She is unpredictable, playful, and loves making fun of the adults in her life who she perceives to be annoying and unreasonable.

Written during a critical point in world history (WWII) and the women's movement, the Pippi Longstocking series smashes the patriarchy even by today's standards. The series was first published in 1945, a turning point in feminism in which women were liberated from the home to take up jobs and roles historically held by men who had gone to war. Pippi embodies the spirit of these new modern women and girls who were all about getting the job done.

The series broke a lot of stereotypes about "good girls" (who usually grow up to be "good women"). Pippi was strong, confident, uninhibited, she stood up for the poor and downtrodden, and questioned authority and society. She directly contrasted her prim-and-proper friend Annika who was considered the "ideal girl." Of course, it is this contrast that makes Pippi stand out and be even more striking! Pippi is such an exciting character because she breaks with conventional ideas about how girls should behave and makes fun of absurd gender roles in the process.

ABOUT THE AUTHOR

Astrid Lindgren was born in Sweden in 1907 and was as unusual and as bold as Pippi herself. She experienced WWII from Stockholm and her joyful tales of Pippi offered many in Europe a much-needed distraction from the horrors of war. As Astrid put it herself, "If I have managed to brighten up even one gloomy childhood—then I'm satisfied."

Astrid was an astute businesswoman, humanist, and campaigner for children. Her funeral (on International Women's Day no less, after passing at the ripe old age of ninety-four) was attended by the Swedish royal family and Prime Minister. Thousands of others lined the streets to say their goodbyes to this much beloved author. Through her work she traveled the world signing books and giving speeches promoting literature, peace, and women's rights. She also campaigned for fairer taxes and laws against animal cruelty and child pornography.

FACTS

- Astrid's books have been translated in over seventy languages.
- According to Pippi herself, her full name is Pippilotta Delicatessa Windowshade Mackrelmint Ephraim's Daughter Longstocking.
- It was Astrid's daughter who came up with the name Pippi Longstocking when she was sick and asked her mom to come up with a story for her character.
- Astrid started telling Pippi stories for her daughter back in 1941 but it wasn't until a few years later, after her daughter fell on ice and became immobile, that Astrid put pen to paper and wrote them down.

MATILDA WORMWOOD,
MATILDA

"She longed to do something truly heroic."

LESSON LEARNED
Being unique can be truly magical.

PLOT
Matilda is a gifted little girl growing up in a family with a cruel mother and father who treat her with disdain. Unlike the rest of her family, Matilda loves reading and has a thirst for knowledge. When Matilda starts school, she meets Miss Honey, her teacher who is sweet and encouraging. She gives Matilda the affection she has missed her whole life. Unfortunately, the school is run by a terrible head teacher, Miss Trunchbull, who is a massive bully and makes the schoolchildren's lives miserable. When Matilda's friends are berated by Miss Trunchbull, Matilda uses her newly discovered telekinesis powers to reap revenge! The story ends with Matilda finally finding a happy home with Miss Honey, proving that women don't need men to find their happily ever after.

FEMINIST ICON
Roald Dahl has written many children's books, but none have quite as strong a female character as the precocious five-and-a-half-year-old Matilda. By walking to the library and teaching herself to read, Matilda worked hard to transform her tragic home life into something more meaningful and magical. While sexism isn't a prominent theme in the novel, there is definitely some discrimination against women. When Matilda first begins to show signs of intellect by speaking at the age of one and a half, she is told that "little girls should be seen and not heard."

Instead of trying to reason with a society that belittles her, Matilda takes it upon herself to learn as much as she can so that one day, she is able to improve the system. She proved that girls are smart and serve a greater purpose. If that's not inspirational enough, her intelligence eventually leads to telekinesis, which allows her to stand up for herself and her classmates against inhumane treatment from her parents and the evil Miss Trunchbull.

Matilda's keen mind and great imagination better her life and the lives of those around her, making her a trailblazer for young feminists. Just like the real-life feminist heroes Malala Yousafzai and Greta Thunberg,

Matilda is proof that nothing is going to stop young girls from changing the world for the better.

ABOUT THE AUTHOR

Roald Dahl is a celebrated children's author who is honored with his own national day: September 13. He wrote more than twenty children's books, including *The Twits*, *The BFG*, and *Charlie and the Chocolate Factory*, many of which have been turned into blockbuster films.

Many might not know that Roald Dahl was a British fighter pilot and spy during WWII, passing intelligence to MI6 from Washington (less James Bond and more *James and the Giant Peach*!). That said, Dahl *did* have a part to play in the James Bond franchise. Working alongside creator Ian Fleming, Dahl wrote the storyline for the fifth James Bond movie, *You Only Live Twice* (1967). In his personal life, Dahl had a reputation of being a womanizer, bully, and a bigot. Perhaps he was not quite the saint he is often made out to be.

FACTS

- When Roald Dahl died in 1990, he was buried with some of his favorite things, including a power drill, chocolate, a snooker cue, and, like any writer worth his salt, an HB pencil.
- He created over five hundred new words such as Oompa-Loompa and scrumdiddlyumptious in *Charlie and the Chocolate Factory*.
- It was through his time as a pilot that Dahl wrote his first children's book, *Gremlins*, about gremlins who caused all sorts of mechanical problems on airplanes. It was this book that inspired Spielberg's blockbuster movie of the same name.
- Dahl also adapted Ian Fleming's story *Chitty Chitty Bang Bang* for the big screen.

ANNE SHIRLEY,
ANNE OF GREEN GABLES SERIES

"Oh, it's delightful to have ambitions. I'm so glad I have such a lot. And there never seems to be any end to them—that's the best of it."

LESSON LEARNED

Be the person you want to be, regardless of what you've been through. Your past has no claim on your future.

PLOT

Anne (with an "e") Shirley is a smart, willful, and imaginative young eleven-year-old orphan girl growing up in the 1880s in rural Canada. It is here where siblings Marilla and Matthew Cuthbert are looking to adopt a young boy to help on their family farm, Green Gables. By mistake, Anne is sent to the siblings and, although Matthew is instantly taken with her, she has to convince Marilla that she is better than any boy would have been. Anne's cheerful and highly imaginative nature allows her to charm those around her (and have a few adventures along the way, too!).

FEMINIST ICON

Spunky, imaginative, and irresistibly fun, Anne doesn't let her tragic background stop her from being her best and most optimistic self. When Anne shows up to the Cuthbert's farm, she realizes that they were hoping for a boy and not a girl (which probably isn't too far off from how society felt about all children during this time—I'm imagining gender reveals for girls accompanied by sad trombones). Rather than feeling defeated or inadequate, she continues to stay positive and fight to prove her value. Many trials ensue and she has a lot to overcome and learn about herself and others.

Because Anne had a chaotic start to life, she never learned the proper etiquette that a girl of her era would be expected to abide by. As a result, she unintentionally rebels against the status quo of her sleepy provincial town, but Anne doesn't mind that people find her odd, nor is she discouraged by her blunders. Rather, she persistently learns from her mistakes and always remains true to herself. With a little hard work and a whole lot of imagination, Anne becomes an accomplished student, teacher, and writer, eventually endearing herself to everyone who knows her.

Montgomery, just like Anne, was a strong feminist of her time. Her family considered writing to be a waste of time, especially for a woman, so she worked in secret, even going so far as to smuggle candles into her room so that she could write at night while her family slept. Montgomery would often write under pen names such as "Joyce Cavendish" so that her family and friends wouldn't discover her writing and so that readers couldn't tell what gender she was.

ABOUT THE AUTHOR

Lucy-Maud Montgomery, born in 1874, was a Canadian author who wrote over five hundred short stories, twenty novels, and two poetry collections, many of which are still read around the world. As a young woman, Montgomery had a very spicy love life! She turned down two proposals before getting engaged to her second cousin, Edwin Simpson. However, soon after, she realized she didn't love Simpson and couldn't marry him. Meanwhile, she fell in love with Herman Leard, a farmer's son. Though Montgomery felt strongly about Leard, he didn't have the intelligence she wanted in a partner. It all came to a head when both men visited her at the same time. She wrote in her journal: "There I was under the same roof with two men, one of whom I loved and could never marry, the other whom I had promised to marry, but could never love!" Needless to say, neither relationship lasted.

FACTS

- In 1905, Montgomery sent the manuscript of *Anne of Green Gables* to several publishers who all rejected the book. Discouraged, she stuck the novel in a hatbox. Two years later, she came upon it, polished it up, and sent it out again. This time, L.C. Page & Company in Boston agreed to publish the novel.
- *Anne of Green Gables* was an immediate success, selling nineteen thousand copies in its first five months.
- Besides *Anne of Green Gables*, Montgomery's bestsellers included *Emily of New Moon*, *Jane of Lantern Hill*, *Pat of Silver Bush*, *The Blue Castle*, and *The Story Girl*.
- Mindy Kaling is a self-professed *Anne of Green Gables* fan and said it's one of her favorite books in an interview with the *L.A.*

Times. "Living on Prince Edward Island would be so badass," she added.

- The last book in the Anne of Green Gables series, *The Blythes Are Quoted*, was published posthumously in 2009. Her obituary in *The New York Times* stated that the manuscript had been delivered to her publisher the day before her death.

PART TWO

Teenage Drama Queens

Stories are windows into other people's worlds. As a teenager, when your world is rapidly expanding, the stories you consume and the fictional characters you meet can often form many of your beliefs and world views.

Luckily, more and more of young adult (YA) fiction now features strong female characters. The reasoning behind this is simple: teenage girls read a lot more than boys. In 2009, a global study[*] of the academic performance of fifteen-year-olds found that, in all but one of the sixty-five participating countries, more girls than boys said they read for pleasure. Only about half of boys said they read for enjoyment, compared to roughly three-quarters of girls. As a result, you don't need to look hard in YA fiction to find excellent female role models.

This section is dedicated to strong teenage protagonists. They are not only going through all the awkwardness that comes with puberty and growing up but are also changing the world for the better, like Persephone and Starr, embracing what makes them unique, like Wednesday, or simply being a total badass and saving the day, like Nancy and Buffy.

---

[*]   OECD (2010), *PISA 2009 at a Glance*, OECD Publishing, Paris, France

WEDNESDAY ADDAMS,
THE ADDAMS FAMILY

"So, I'll have to prove my power."

LESSON LEARNED

Protect your energy and smile only when you want to.

PLOT

Starting in 1938, Gomez, Morticia, Uncle Fester, Lurch, Grandmama, Wednesday, Pugsley, and Thing appeared in *The New Yorker* in a series of cartoons by Charles Addams. The macabre family have gone on to feature in sitcoms, Saturday morning TV cartoons, and feature-length films.

FEMINIST ICON

The Addams Family was one of the most modern and feminist families you could find in television and film in the early nineties. All the women of the Addams family are independent with their own style and strong narrative. At the center is Wednesday Addams, a feminist role model for the ages. She's brainy, rebellious, and wise beyond her years by uniting the marginalized in an act of protest against conformism, classism, racism, ableism, and sexism. Wednesday Addams flies in the face of convention, refusing to conform to what society considers "normal." Her refusal to play by the rules and really-don't-care attitude is what most appealed to me growing up. She's a total rebel.

In a world where women are constantly told to smile, Wednesday is an excellent reminder that we are not here to pacify others. She is the queen of the resting bitch face and *owns* it. Like many women, there have been occasions where I have succumbed to societal pressures and flashed my pearly whites upon command. Well, no more! Take inspiration from Wednesday herself who only smiles when she feels like it (most likely at the suffering of someone who wronged her).

Wednesday is not afraid to get political and push against a false narrative. In the *Addams Family Values* film, Wednesday pretends to play the insipid Pocahontas as written for a play about the first Thanksgiving but goes off-script with a feminist monologue calling out the ill-treatment of the pilgrims against the Native Americans.

Wednesday is also independently minded when it comes to dating. She takes full control of the budding romance with her campmate Joel. It is she who initiates the relationship and asks him to go with her to Uncle

Fester's wedding (granted, she won't refer to it as a date) before Joel has the guts to ask her out. She teaches us to be strong-willed and go after what you want rather than always waiting for the boys to make the first move.

Wednesday is peculiar, spooky, and even terrifying on occasion. Her dark qualities oppose the simplicity of the sunny side of feminism we often see in the media. For all these reasons, Wednesday is a feminist powerhouse!

FACTS

- There was an actor rebellion on set of *The Addams Family* led by ten-year-old Christina Ricci (who played Wednesday in the 1991 hit film). Actors were not happy with some of the script and plot development and nominated Christina to give an impassioned plea to Barry Sonnenfeld, the director. The plot changed for the better, thanks to Christina.
- Sonnenfeld notes the unrealistic body shape that Morticia Addams has, telling *Entertainment Weekly*, "Morticia has a shape only a cartoonist can draw . . . so we lashed Anjelica Huston into a metal corset that created this hips-and-waist thing I've never seen any woman have in reality."
- The house we see in the opening credits was a real house located at 21 Chester Place in Los Angeles, California. To make it a little spookier, special effects technicians added a third floor with a tower. However, the house has since been demolished.

PERSEPHONE HADLEY, NOUGHTS & CROSSES SERIES

"What was it about the differences in others that scared some people so much?"

LESSON LEARNED
Privilege is invisible to those who have it.

PLOT
This young adult fiction series is set in a dystopian world where those with black skin are called "Crosses" and are considered the ruling class while those with white skin, known as "Noughts," are considered second-class citizens with low-paid menial jobs and fewer rights. The story revolves around the forbidden love between Persephone (Sephy) and Callum, a black girl and white boy. The series tackles huge issues such as race, discrimination, and prejudice.

FEMINIST ICON
I first came across the Noughts & Crosses series as a young teenager. The first book in the series was included on my summer reading list, and to this day it is one of the most memorable books I have ever read. It was reading this series that marked the start of my education about our society's racial, political, and social injustices. These themes are of course enormous, but *Noughts & Crosses* so elegantly makes them understandable to all. For me personally, growing up in the countryside where everyone looked like me, the book opened my mind to my own privilege and got me thinking and asking questions about race for the first time.

This series is written through the perspective of different characters, offering the reader a better understanding of how each character feels and thinks. Persephone (Sephy) provides the Crosses perspective and is the daughter of a powerful politician. Callum, the son of Sephy's nanny, provides the Noughts perspective. The series develops around the pair's friendship which grows into romance. However, the couple face many challenges coming from all areas of society.

Sephy at times comes across as naïve and self-centered due to having grown up in a very privileged environment, but she has many wonderful qualities which make her a strong female protagonist. From a young age she has seen past the racist world she lives in and has also come from a place of good intentions. Time after time she speaks out about injustice and tries to take actions to create a better world. Even when an act as simple as choosing to eat lunch with the Noughts at school leads to three older Crosses beating her up for not knowing her place in society, Sephy is brave and continues to stand up for what she believes is morally right, inspiring young girls everywhere.

ABOUT THE AUTHOR

Malorie Blackman is a British author with an impressive body of work comprising more than sixty books. She has received many accolades during her career and in 2008 was awarded an OBE (a British order of chivalry) by the Queen for service to children's literature.

FACTS

- There are six books in the series: *Noughts & Crosses*, *Knife Edge*, *Checkmate*, *Double Cross*, *Crossfire*, and *Endgame*.
- Malorie Blackman was responsible for the groundbreaking 2019 Rosa Parks episode of *Doctor Who*, along with executive producer Chris Chibnail.
- The working title of the book series was *Snakes and Ladders*. However, Blackman admitted the name never sounded right to her.
- *Noughts & Crosses* was adapted into a BBC TV series in 2020.

NANCY DREW,
NANCY DREW SERIES

*"I have solved some mysteries, I'll admit, and I enjoy it,
but I'm sure there are many girls who could do the same."*

LESSON LEARNED
Don't just talk about something . . . live it! Use every resource you have and put your ambition to work.

PLOT
Nancy Drew is a sixteen-year-old amateur sleuth who lives in the town of River Heights, Illinois, with her father, attorney Carson Drew, and their housekeeper, Hannah Gruen. After losing her mother at a young age, Nancy spends her time solving mysteries, some of which she stumbles upon and some of which are her father's cases. Nancy is often assisted on her adventures by her two closest friends: Bess, who is described as delicate and exhibits attributes associated with traditional femininity, and George, who is a fun-loving tomboy. Her only distraction from the pursuit of the truth is that of her crush, Ned Nickerson. Not exactly "girl power" but, then again, what sixteen-year-old girl has not made the mistake of placing romance above friendship on occasion?

FEMINIST ICON
The Nancy Drew series was published by Edward Stratemeyer under the pen name Carolyn Keene and is attributed to a team of ghost writers, the most prolific of whom was Mildred Wirt Benson. She penned twenty-three stories over twenty-four years. Nancy Drew embodies the feminist paradox, with some arguing that she represents a mythic hero, an expression of wish fulfillment, or an embodiment of contradictory ideas about femininity. For me, she serves as a wonderful role model. I would argue that it takes a strong individual to come across a hidden tunnel or forbidden staircase and not run as fast as possible in the opposite direction! Nancy Drew always faces her fears head-on and doesn't let anyone hold her back. She perfectly illustrates how sometimes it is necessary to lean into discomfort in order to succeed.

Because of her intense interest in her surroundings, she is a seriously well-rounded person, especially at such a young age. Aside from being a detective as well as a high school student, she also has time to practice horseback riding, cooking, and sewing (while always looking trendy in her penny loafers). Above all, Nancy Drew is confident, independent, and capable of solving problems on her own. But let's not forget that Nancy Drew is accomplished because she chooses to be, not because she just is. She's far too ambitious to let opportunities idly pass her by.

ABOUT THE AUTHOR

Edward Stratemeyer was born to middle-class German immigrants in New Jersey in 1862. He found tremendous success as a writer and is considered one of the most prolific writers of all time, producing thirteen hundred books and selling more than five hundred million copies. Edward Stratemeyer gave life to literature's most famous teenager just before his sixty-seventh birthday. Books featuring his final and probably best-known creation became widely available on April 28, 1930, and Stratemeyer died just twelve days later. His two daughters Harriet and Edna took over the family business and, seeing the potential in Nancy Drew, went on to create many more stories.

FACTS

- Many of today's most powerful and influential women have cited Nancy Drew as an inspirational character, including Ruth Bader Ginsburg, Sandra Day O'Connor, Laura Bush, Hillary Clinton, and Barbara Walters.
- It is estimated that eighty million copies of Nancy Drew have been sold worldwide in forty-five different languages.
- One of the names originally suggested for Nancy Drew was Stella Strong.
- Nancy Drew has featured in over five hundred books and several feature films. She has recently been given a twenty-first-century technological makeover. (Well, she has been solving crimes for more than ninety years . . . it's about time her father trusted her with her own phone!)

STARR CARTER,
THE HATE U GIVE

"That's why people are speaking out, huh?
Because it won't change if we don't say something."

LESSON LEARNED

Use your voice to spark change and call out injustice.

PLOT

The Hate U Give follows Starr Carter, a sixteen-year-old black girl whose world changes after she witnesses the shooting of her best friend. The book hit *The New York Times* bestseller list, inspired hundreds of young activists, and believe it or not, was banned by some authorities and institutions across the United States. At its core, the book follows the interaction of Starr as she inhabits two parallel worlds: the poor, mostly black neighborhood where she lives, and the wealthy, mostly white prep school that she attends.

FEMINIST ICON

Starr is an amazing heroine with a powerful story. She is a shy girl that experiences the exhausting pressures that come from being a young black woman existing in two very different worlds, experiencing both white privilege and racism. This ability of having to maintain two versions of herself results in Starr never truly feeling like herself. After Starr witnessed the murder of an innocent friend, Khalil, she experiences the injustice and prejudice that exists in modern-day America. It is a story that has been played out across the country, that when girls and women attempt to seek justice and tell authorities about what they witnessed or experienced, they are often dismissed or not believed.

Starr attempts to repress the memories that torment her and instead invests in becoming the kind of person her friends and family need her to be. She's the dutiful daughter for her parents, never letting her grades drop. For her white friends, she's the nonthreatening girl who allows them to fantasize about being black, spitting slang and rapping lyrics from the latest rap song, while carrying none of its burden. While for her friends she grew up with, she is the loyal girl they have always known. Despite performing these different parts, the reality is that Starr never allows herself to build a cohesive sense of self, and she eventually begins to unravel.

Her emotions come to a boiling point when her white friends and the media insist that Khalil was a thug and drug dealer who would have been killed anyway because he (and the hairbrush he was carrying, which the officer mistook for a gun) was a public threat. Starr recognizes the ignorance of those around her and how she will never fit in because there will always be someone perceiving her as a criminal because of her race. After the officer involved in the shooting is acquitted of shooting Khalil, Starr arms herself with her most powerful weapon, her voice, and calls out the injustice she has witnessed. Starr is an inspiration, and her strength and courage to speak out makes her a feminist icon we should all learn from.

ABOUT THE AUTHOR

Author Angie Thomas was inspired to write her debut novel, *The Hate U Give*, after the tragic death of Oscar Grant, an innocent twenty-two-year-old father who was fatally shot by a white officer in Oakland, California, while trying to break up a fight he had witnessed outside his barbershop. The officer claimed he thought Oscar was reaching for a concealed firearm, but he was unarmed. The story that started out as a passion project for Thomas went on to receive numerous accolades, including the Coretta Scott King Book Award (2018) and nominations for the Los Angeles Times Book Prize (2018), the Kirkus Prize (2017), and many more. It has been translated into over twenty-five languages.

FACTS

- The title *The Hate U Give* was inspired by lyrics of Thomas's favorite rapper Tupac Shakur.
- The book and film version of *The Hate U Give* have different endings.
- Amandla Stenberg, who plays Starr Carter in the film adaptation, is also a famous civil rights activist. She's played other strong female characters such as Rue in *The Hunger Games* and Ruby in *The Darkest Minds*.
- *The Hate U Give* spent fifty weeks at the top *of The New York Times* bestseller list after its publication in February 2017.

BUFFY SUMMERS,
BUFFY THE VAMPIRE SLAYER

"Alright, yes, [I want to] date and shop and hang out and go to school and save the world from unspeakable demons. You know, I wanna do girlie stuff."

LESSON LEARNED
How to slay the Buffy way!

PLOT
Buffy the Vampire Slayer was a cult TV series that aired between 1997–2003 and revolved around the adventures of Buffy Summers, a popular teenage girl who, as the chosen slayer of her generation, has to battle vampires, demons, and other supernatural forces of evil. She is aided by her loyal "Scooby gang," including Willow, Xander, and Giles, who help and support her along the way.

FEMINIST ICON
For me, growing up in the nineties, Buffy was the first female protagonist who really tackled the issues of female empowerment. She was the poster girl for third-wave feminism in popular culture. Over the course of seven seasons, Buffy experienced every single problem imaginable from problematic love triangles to the perils of low-paid employment (and the big stuff like saving the whole goddamn world!). And although my own dramas may not have been on the same scale, I often found myself thinking *what would Buffy do?*

Buffy is an amazing example of trusting your female intuition. She stays cool, calm, and collected in even the most dangerous situations. She trusts her instincts and knows exactly what to do to defeat the demons she encounters. Buffy taught me to have faith in myself and trust my gut, something so important for women and girls to know. When a situation does not feel right, it probably isn't, so be like Buffy and fight those demons (not necessarily with a stake through the heart but call them out for what they are).

Buffy also shows us what it means to be a good friend. Even when Willow turns to the dark side and threatens to end humanity, Buffy does not give up on her. Just like Buffy and Willow, the importance of cultivating female friendships cannot be underestimated. Strong women support strong women. Even if that means having difficult conversations and sometimes saying things that are hard to hear, the Willows of the world will understand it comes from a place of love and will love you back harder and fiercer.

The final lesson Buffy teaches us is that being the boss does not make you a bitch. Buffy is the ultimate Boss Babe. She knows it, the whole gang knows it, even the demons know it! Buffy never apologizes for making the hard decisions or telling people what to do. If she isn't being heard, she shouts; if someone is in her way, she round-house kicks them out of it. Buffy epitomized female empowerment during a time when pop stars were asking men to "hit me baby one more time." She was girl power before girl power was cool. She created the canon for ass-kicking heroines to save the day.

FACTS

- The show's creator, Joss Whedon, wanted to take the classic "victim" type, a young, blonde, seemingly harmless girl from Hollywood horror movies, and turn her into the hero.
- *Buffy* was the first show to use the word "Google" as a verb on TV. In the fourth episode of Season 7, Willow asks Buffy, "Have you Googled her yet?"
- Actresses Sarah Michelle Gellar (Buffy) and Allison Hannigan (Willow) are the only cast members to appear in all 144 episodes of the show.
- The high school that was used as the filming location for Sunnydale High can also be seen in *Beverly Hills 90210, The Secret Life of the American Teenager, She's All That, Bruce Almighty*, and *Not Another Teen Movie*.
- Some of the cast had trouble with the fast-paced, Valley girl-esque dialogue, particularly Sarah Michelle Gellar, who grew up in New York and didn't understand a lot of the California slang.

CARMEN LOWELL, *THE SISTERHOOD OF THE TRAVELING PANTS*

*"Sometimes it seems like we're so close we form
one single complete person rather than four separate ones."*

LESSON LEARNED

The power and love of female friendships.

PLOT

Whether it was through the book series by beloved young adult (YA) novelist Ann Brasheres or the Hollywood blockbusters, however you discovered *The Sisterhood of the Traveling Pants* you were in for a treat. The story follows the adventures of four best friends, Lena Kaligaris, Tibby Rollins, Bridget Vreeland, and Carmen Lowell, who will be spending their first summer apart. When a magical pair of jeans comes into their lives, their summer is turned upside down.

FEMINISTS ICON

What makes this story so captivating is the enduring strength of female friendship in the sometimes turbulent teenage years. When I first read the series, I was the same age as these girls. Reading about their one crazy summer (and one extremely flattering pair of jeans) taught me some huge lessons about what it means to be a true BFF.

There is not a single group of teenage girls that hasn't occasionally had a bit of drama to deal with, whether it is liking the same guy, tagging a friend in an unflattering photo, or falling out over a simple miscommunication. Navigating teenage friendships can be a minefield where one wrong move can see you ostracized, blocked, and unfriended.

However, this was not the case with Bridget, Tibby, Carmen, and Lena. The girls couldn't be more different. Bridget is sporty and tough, Tibby is outspoken and creative, Carmen is fiery and sensitive, and Lena is shy and reserved. Carmen even muses that the four of them together make up one whole person together. But just because friends aren't exactly like each other doesn't mean they can't be close. *The Sisterhood of the Traveling Pants* is all about the strength in the sisterhood, and even if you don't have a pair of amazing magical pants in your friendship, the real magic is in the support you have for one another.

FACTS

- The first film was released in 2005 and earned $42 million worldwide. The sequel followed in 2008 and earned $44.4 million worldwide.
- America Ferrera, who plays Carmen in both films, didn't read the script when it was sent to her. It sat on her desk for a year until her mother read it and convinced her to take the part.
- In the movie, the brand of jeans used was Levi's, although it's likely that each actress had her own pair to ensure the jeans really did fit properly. Sadly, the perfect jeans described in the film don't exist in real life.
- All the actresses had to learn new skills for the movie. Ferrera had to learn to play tennis, Alexis Bledel (Lena) had to be taught how to ride a Vespa, and Blake Lively (Bridget) had to undergo major soccer training before filming began.
- In the film, Blake's dad is played by her real father, Ernie Lively, who is an actor in his own right, known for projects like *The Dukes of Hazzard* series, *Quantum Leap*, and *Passenger 57*.

PART THREE
Magical Maidens

From Cleopatra to Clinton, powerful women throughout history have often been branded with the term "witch" as a way to demonize and discredit them. This insult was historically a death sentence, as was the case for Joan of Arc and Anne Boleyn, who both grew too powerful for the society they lived in. When the British could not defeat Joan of Arc on the battlefield, they used the evidence of her superior intelligence and bravery as a sign of witchcraft and burned her at the stake. When the relationship between King Henry VIII and his second wife Anne Boleyn soured and he was already eyeing a third wife, he accused Boleyn of witchcraft and had her beheaded.

Women in politics today still find the term witch being hurled at them. Julia Gillard, first female prime minister of Australia, often received taunts of "ditch the witch" from protesters. When British prime minister Margaret Thatcher died in 2013, the song "Ding-Dong! The Witch Is Dead" from *The Wizard of Oz* jumped up on the music charts. Nancy Pelosi, minority speaker of the US House of Representatives, has faced similar witch-related insults, and when former British prime minster Theresa May was filmed laughing loudly, her "witch's cackle" quickly went viral. No matter what side of the political spectrum women are on, they can't escape the toil and trouble that comes with the witch slur (something their male counterparts don't often have to contend with).

The infamous witch trials which spread like wildfire throughout the ages in Europe and America saw thousands of innocent women murdered as a way to try and control women and stop them from becoming powerful.

But the concept of women being called witches predates the Christian era and can be traced all over the world in different cultures.

Despite centuries of societies using witch as an insult to keep women down, witches are more prevalent in popular culture than ever. With books and shows like *The Discovery of Witches* and *WandaVision*, we seem bewitched by witches. That's why this next section features some of the most enchanting magical maidens who taught me to embrace my own powers.

HERMIONE GRANGER, HARRY POTTER FRANCHISE

*"Books! And cleverness! There are more important things—
friendship and bravery . . . "*

LESSON LEARNED

Intelligence, combined with compassion and loyalty, is an invaluable tool in accomplishing great things.

PLOT

The Harry Potter franchise is a story about a boy who, on his eleventh birthday, finds out that he is the orphan of a witch and wizard and possesses magical powers of his own. He is invited to become a student at the magical school Hogwarts, where he meets new friends, Ron Weasley and Hermione Granger, and learns more about the wizarding world and the dangers it holds for him.

FEMINIST ICON

Not only is Hermione Granger the smartest witch of her age, she's also a badass feminist and one of the most beloved heroines in the Harry Potter Universe. Logical, witty, and extremely knowledgeable, Hermione isn't bothered by her reputation as a know-it-all. Instead, she embraces her perfectionism and excels in every class.

Hermione had a different upbringing than many children at Hogwarts, coming from "muggle-born" (non-magical) parents, so she taught herself much of the practical knowledge of the wizarding world in order to fit in with her new classmates. Her muggle-born status, and the fact that she is a successful witch, makes her a target for many who believe that only "pure-blood" witches should be able to succeed and thrive. She is frequently targeted by pure-blood Draco Malfoy but when matters came to a boiling point Hermione fights her own battle to give Malfoy what he deserves!

Hermione's passionate nature and strict moral code of ethics lead her to spearhead the group S.P.E.W. (Society for the Promotion of Elfish Welfare), which protests the mistreatment and slavery of house elves. Even though her friends are not fully on board with her cause, Hermione does not give up and she keeps on fighting for what she believes in. For all these reasons and more, Hermione is a great role model and wand-waving feminist icon!

ABOUT THE AUTHOR

J. K. Rowling is a British author and philanthropist born in Gloucestershire, England, in 1965. Rowling came up with the concept of Harry Potter on a delayed train from Manchester to London in 1990. She has lived a rags-to-riches life in which she progressed from living on benefits to being named the world's first billionaire author by *Forbes*. She has given much of her wealth away to charities such as her own charity Lumos, which promotes the end of the institutionalization of children worldwide.

FACTS

- The Harry Potter book series has sold over 500 million copies and is the bestselling book series in history.
- J. K. Rowling's actual name is Joanne Rowling. It was a strategic decision on the part of her publisher to release the book under a gender-neutral name to make male readers more inclined to pick it up!
- The Harry Potter manuscript was rejected twelve times by different publishing houses before it was accepted by Bloomsbury in 1997.
- Unsurprisingly, Emma Watson, who brought Hermione to life in the movie franchise, is also a feminist icon in her own right. She delivered a rousing speech at the United Nations in 2014 calling everyone to declare themselves as feminists!
- Dementors in the books are based on Rowling's own struggle with depression after her mother died.

MEG MURRY,
A WRINKLE IN TIME

*"Maybe I don't like being different . . .
but I don't want to be like everybody else, either."*

LESSON LEARNED

Asking for help is not a sign of weakness, but rather a sign that you want to remain strong.

PLOT

A Wrinkle in Time is the story of Meg Murry, a high school girl who is transported on an adventure through time and space with her younger brother Charles Wallace and her friend Calvin O'Keefe to rescue her father, a gifted scientist, from the evil forces that hold him prisoner on another planet. Throughout their quest, the children are forced to confront powerful evil forces while being guided by three celestial beings, Mrs. Which, Mrs. Who, and Mrs. Whatsit. Originally released back in 1962, *A Wrinkle in Time* is regarded as one of the most important sci-fi novels and has since been adapted for the big screen.

FEMINIST ICON

Meg Murry was an easy-to-relate-to young heroine in a male-heavy genre of sci-fi adventure fiction. Never mind that I personally had never been tasked by celestial beings to undertake an interstellar quest defying the rules of physics to save my missing father. No, what mattered was that there was finally a character I could identify with: a bold misfit (with a slightly annoying younger brother!). And unlike the other kick-ass females in sci-fi, Meg was not gifted with any supernatural powers. She was not the chosen one. She was a normal human teenage girl just like I was, and she still got to be the lead of her own amazing adventure. She is not like many of the male protagonists from other sci-fi stories who fight or blast their way out of a problem. Meg uses her inner strength and courage to help save the day.

Meg's passion for STEM (Science, Technology, Engineering, and Math), which has historically been associated with males, makes her an inspiring female role model. Meg uses her considerable intellect to conquer the heartless, power-hungry darkness referred to as "IT." And while she draws on her own human strengths to do so, she also heeds the erudite wit of her cosmic goddess guides: the unearthly Mrs. Whatsit, Mrs. Who, and Mrs. Which, teaching us that accepting help from others is a sign of strength rather than a sign of weakness. No one is expected to be born with all the answers. We will all need some guidance throughout our lives, whether that is from the likes of three goddesses or from those closer to home.

ABOUT THE AUTHOR

Author Madeleine L'Engle is a revered writer of nonfiction. However, she nearly gave up writing at age forty before finding herself inspired to start writing *A Wrinkle in Time* while on a ten-week cross-country camping trip. L'Engle was heavily influenced by her interest in quantum physics and theories on cosmology. The idea that religion, science, and magic are different aspects of a single reality and should not be thought of as conflicting is a recurring theme in her work. Oddly enough, *A Wrinkle in Time* has been accused of being both too religious and anti-Christian. As a result, it is one of the most frequently banned books.

FACTS

- L'Engle often compared her young heroine, Meg Murry, to her childhood self—gangly, awkward, and a poor student.
- L'Engle weathered twenty-six rejections before Farrar, Straus and Giroux finally took a chance on *A Wrinkle in Time*.
- *A Wrinkle in Time* is the first novel in the Time Quintet series, which follows the Murry family's continuing battle with evil forces for thirty years.
- The Disney movie features Oprah Winfrey, Reese Witherspoon, and Mindy Kaling as celestial guides Mrs. Which, Mrs. Whatsit, and Mrs. Who. All three of these fantastic actresses are feminist icons in their own right!

SABRINA SPELLMAN,
SABRINA THE TEENAGE WITCH

*"What's the matter? I have to be a witch, I have to be a mortal,
I have to be a teenager, and I have to be a girl all at the same time.
That's what's the matter."*

LESSON LEARNED
There is no quick fix to life's problems—not even magic.

PLOT
Sabrina the Teenage Witch chronicles the adventures and many dramas of Sabrina Spellman, a teenager who discovers that she is a witch on her sixteenth birthday. Her two aunts, Hilda and Zelda, as well as the family pet cat, Salem, offer guidance on how to control her newly discovered magical powers. The show was given a gothic reboot with a lot more hex appeal thanks to Netflix with *The Chilling Adventures of Sabrina*. Although the plot is similar, it's a whole different broomstick, with much of the lighthearted comedy removed.

FEMINIST ICON
When I was growing up, I couldn't get enough of *Sabrina the Teenage Witch*. I was obsessed and spent hours fantasizing about having the ability to do magic at just the point of a finger! At its essence the show is a sitcom filled with normal teenage drama: friends falling out, finding a date to the homecoming dance, and dealing with the school bully. But Sabrina is different from the typical leads you find on other shows of the time like *90120* and *One Tree Hill*. She was funny. Sabrina was confident and hilarious and one of my first introductions to a strong woman in comedy.

Sabrina also shows the struggle of having to juggle opposing worlds. For her, this is finding the balance between the supernatural world and that of being a teenage girl with high school politics and biology exams to pass. Many people encounter that same pressure of existing between two parallel worlds, not feeling at home in either. This theme was something Sabrina vocalized regularly throughout the series, during a time when few other shows put any importance on mental health issues.

Sabrina the Teenage Witch did not shy away from tackling some of the bigger issues such as institutional sexism and body confidence. Just like many girls struggling through those awkward teenage years, Sabrina faced it all and threw down some life lessons along the way. In one fateful episode, Sabrina accidentally sent everyone back in time and, while the fashion was fun, this episode highlighted the fact that women did not always have the right to vote or to a higher education. Not many children's shows at the time were commenting on women's suffrage.

At its core the show revolves around three strong women, Sabrina and her two aunts, living together in a nonconventional family and tackling whatever the magical universe throws their way. The only man in the Spellman household was an adorable megalomaniac talking cat, once a warlock who had been cursed to a feline body as penance for a failed attempt to take over the world. All men came secondary to the bold Spellman women (even dreamy Harvey). Sabrina was a fantastic role model with her sharp wit, independence, and comic genius.

FACTS

- Sabrina the Teenage Witch first appeared in Archie Comics.
- Despite the show's age-driven title, when the show premiered in 1996, Melissa Joan Hart was twenty years old. When the show ended, the actress was almost thirty.
- Salem the cat was played by three different real cats called Elvis, Witch, and Warlock.
- The show has dozens of cameos, among them Ryan Reynolds, The Backstreet Boys, NSYNC, Britney Spears, RuPaul, and even Bryan Cranston.

ZÉLIE ADEBOLA,
CHILDREN OF BLOOD AND BONE

"I won't let your ignorance silence my pain."

LESSON LEARNED

Make sure that the walls you build to protect yourself don't become a prison.

PLOT

Children of Blood and Bone is set in the kingdom of Orïsha, a realm inhabited by two distinct people: Divîners (who can become magical Maji and are known for their white hair) and Kosidán (non-magical people). The story follows a young woman named Zélie Adebola who witnesses the death of her mother and other Divîners under the order of a ruthless king who learns of a way to separate the magic from the Maji. Since this time, Divîners have been severely oppressed and forced to survive on the edge of society. Zélie sets out to restore magic to her land, teaming up with a rogue princess to outwit and outrun the crowned prince while avoiding prowling snow "leoponaires" and vengeful spirits.

FEMINIST ICON

Children of Blood and Bone features real West African mythology, which is beautifully woven into the story, adding real color to the fantasy world of Orïsha. The story follows Zélie, a powerful, strong-willed protagonist. Her story is one that rings true for any marginalized person or group who has risen up in protest against their suppressors.

Throughout the novel, there is a stark divide between the Divîners and Kosidán. Divîners are discriminated against for their silver-white hair that indicates their magical heritage and are not allowed to marry non-magical people or have a position of power. King Saran has spread fear and distrust amongst his kingdom that is aimed at Divîners. This hierarchy not only has parallels with racism and colorism (nobles in Orïsha are lighter skinned than the working class who must toil under the sun to make a living), but also strong themes of gender, with all magic originating from the "Sky Mother," while state power comes from the King.

Zélie is driven and aided by the many women in her life, from her mother who she inherits her powers from, to Mama Agba who teaches her and other Divîners how to fight and defend themselves. But it is Amari, King Saran's own daughter, that Zélie learns the most from. Initially Zélie believes Amari to be nothing more than a spoiled Princess that is weak and defenseless. But the two grow to become like sisters, with Amari's warmth and trust in others showing Zélie that she doesn't need to be so guarded all the time. Lowering your defenses to allow others in can be extremely fulfilling, especially if you have been burned in the past. This is a lesson that I'm sure many of us could learn from.

ABOUT THE AUTHOR

Children of Blood and Bone was Nigerian American novelist Tomi Adeyemi's debut novel and the first in the Legacy of Orïsha series. Adeyemi drew inspiration from novels like *Harry Potter* and *An Ember in the Ashes,* as well as West African mythology and the Yoruba culture and language, to create this magical kingdom. Adeyemi felt compelled by the police shootings of black Americans to write this coming-of-age YA novel. She also wrote this novel for personal reasons. Growing up, Adeyemi experienced the lack of black protagonists in fantasy novels and wanted to remedy this for the next generation.

FACTS

- At just twenty-three years old, Tomi signed a reported seven-figure book deal to bring the story of Orïsha to life. The novel debuted as a *New York Times* bestseller.
- According to Tomi, it took forty-five drafts and eighteen months to complete her book because of her commitment to getting all the small but crucial cultural details in the story right.
- Plans for a film adaptation are underway with Paramount Entertainment, with other big movie producers such as Disney's Lucasfilm also bidding for the rights to the film.
- The names of the villages in the novel are the same as real places in Nigeria.

CIRCE,
GREEK LEGEND

"It is a common saying that women are delicate creatures, flowers, eggs, anything that may be crushed in a moment's carelessness. If I had ever believed it, I no longer did."

LESSON LEARNED
Take the time to learn to love yourself.

PLOT
Circe is credited as the first witch ever recorded in western literature. She first appears in Homer's *Odyssey*. After Odysseus and his crew wash up on her island, exhausted and grieving for the loss of their comrades, they go searching for inhabitants and find a palatial house with tame lions and wolves lolling around in the garden. A shining goddess comes to the door and invites them in. She gives them food and wine which she has enchanted with herbs that turn the men into pigs.

FEMINIST ICON
Madeline Miller's retelling of Circe's story was the feminist epic odyssey that was missing from my life. The story starts with Circe very much an outsider in her own home, bullied and belittled by the more powerful gods and goddesses around her. When she starts to develop her own powers of witchcraft she realizes that she has the potential to be more powerful than she ever thought. From fear of her new growing power, she is exiled from her community and sent to live alone on the island of Aiaia.

At the beginning of Circe's exile, she feels cut off and isolated. This feeling of loneliness and fear of missing out is something we all feel at certain times in our lives. With only her own menagerie of tame lions, bears, and wolves to keep her company, Circe shows the importance of self-care. She uses her time in isolation to learn about herself and become more comfortable in her own skin, growing to appreciate the tranquility that comes from being alone and to stop caring about what everyone else is doing. It takes a very strong woman to be at peace with themselves.

When Circe is repeatedly betrayed and belittled by the men in her life, she's forced again and again to pick up the pieces and soldier on. But unlike the stereotypical view of an evil haggard witch, cackling with warts on her nose, Circe never becomes bitter. She helps those who wronged her (like her nasty sister) and never gives up on love. There are many lessons that can be learned from Circe, but the ultimate takeaway is that it's time we started giving women their voices back!

ABOUT THE AUTHOR

Miller is a renowned American classicist. She is the author of *The Song of Achilles*, which won the 2012 Orange Prize, and *New York Times* bestseller *Circe*, which won the 2019 Indie Choice Award and was shortlisted for the Women's Prize for Fiction. Writing in a genre that Miller refers to as "mythological realism" allows for the retelling of mythological stories in a woman's voice.

FACTS

- Circe is the daughter of the sun god Helios.
- Of the three female sorceresses, Circe, Pasiphae, and Medea, Circe was regarded as the most powerful. She was able to concoct powerful potions and is said to have had the power to hide the sun and moon as she willed.

PART FOUR
Dystopian Divas

The dystopian world has long been a friend of strong women. This is because these post-apocalyptic worlds, where characters are faced with an overwhelming system of oppression with the odds stacked against them, is something women have been facing for centuries!

This genre allows for many of the angers and anxieties surrounding a patriarchal society to be played out to the most extreme, the most well-known being Margaret Atwood's 1985 novel *The Handmaid's Tale*, which has been turned into an award-winning television show. Margaret Atwood's fictional dystopia has inspired real-life political activism. Protesters have dressed as handmaids in the story's iconic red robes and white bonnets to oppose today's policies that restrict women's access to abortion and health care all around the world.

The strong female characters of this genre are inspirational with their rebellious nature and determination to survive. Although the characters are often physically tough, able to wield a sword or a bow and arrow, they are also mentally tough. In a world that does not want them to exist, they persist. Resilience, resourcefulness, and daring to hope in a world that crushes hope—these are the strengths I admire most in this next section's dystopian divas.

JUNE OSBORNE (OFFRED),
THE HANDMAID'S TALE

*"Do not let them grind you down.
You keep your fucking shit together. You fight!"*

LESSON LEARNED
Freedom is something worth fighting for!

PLOT
The novel follows June Osborne, or Offred as she comes to be called, in a dystopian future in which a strict religious regime has taken over the United States and renamed the country the Republic of Gilead. Women are now divided into rigid classes determined by an idiosyncratic interpretation of the bible. Offred becomes a handmaid (a fallen woman who is forced to bear children for righteous couples) and the book follows her treatment under the Gilead regime. The story follows the various means by which the women in Gilead resist and fight for individuality and independence.

FEMINIST ICON
What keeps me up at night is the thought that we are potentially just one bad election away from fiction of *The Handmaid's Tale* becoming reality! In the state of Gilead, women are stripped of all human rights and are subservient to men in every way. They are forbidden from reading, having their own bank accounts, and even choosing their own clothes.

The Handmaid's Tale has become a sort of feminist bible, with June Osborne at the center. She is smart, brave, and although she endures horror after horror, she refuses to let the patriarchy bring her down. She shows us how much power it takes to endure and survive in a world that is against us, and how to be resilient and defy those who try to keep us silent. She is the resistance, she is Mayday! Let "Nolite te bastardes carborundorum" be the rallying cry to keep fighting the good fight against the patriarchy.

ABOUT THE AUTHOR
Margaret Atwood is a Canadian poet, novelist, inventor, and environmental activist who has written many of the twenty-first century's greatest texts of feminist fiction. She has won the Booker prize twice, most recently in 2019 with *The Testament*, the sequel to *The Handmaid's Tale*. Atwood shuns the label of science fiction and instead considers much of her work to be speculative fiction. She explains that "science fiction has monsters

and spaceships; speculative fiction could really happen." If that doesn't scare you into getting political, then you better start practicing living "under his eye!"

FACTS

- The latin phrase "Nolite te bastardes carborundorum" (don't let the bastards grind you down) used in Atwood's novel is totally made up. She jokes she made it up in Latin classes when she was young. The phrase has now taken on a life of its own and is a battle cry for feminists.
- Atwood wrote *The Handmaid's Tale* while living in West Berlin during the Cold War.
- Atwood may be descended from a seventeenth-century Massachusetts woman, Mary Webster, who'd been accused of witchcraft. Mary was sentenced to death by hanging but she survived and earned herself the nickname Half-Hanged Mary.
- Atwood is the first author to contribute to The Future Library Project, which takes one writer's contribution each year for the next hundred years. Unfortunately, we will not know what she has contributed until the year 2114!

HESTER SHAW,
MORTAL ENGINES

*"You aren't a hero and I'm not beautiful and
we probably won't live happily ever after," she said.
"But we're alive and together and we're going to be all right."*

LESSON LEARNED

It's not what's on the outside that counts, but rather who we are on the inside.

PLOT

Mortal Engines is set hundreds of years in the future after a cataclysmic event known as the Sixty Minute War nearly destroyed all of civilization. To survive, towns and cities become motorized in order to move around and compete for resources. Hester Shaw, a young girl driven by the memory of her mother, along with outcast Tom Natsworthy, have to stop those in power from making the same mistakes of the past.

FEMINIST ICON

Mortal Engines is one of the better takes on a post-apocalyptic world. It weaves in many underlying themes of social mobility, privilege, and the devastating consequences of war. Hester is supposed to be horribly disfigured (unlike the small scar in the Hollywood version). In the book series, Hester's disfigurement is described as "Her mouth was wrenched sideways in a permanent sneer, her nose was a smashed stump, and her single eye stared at him out of the wreckage." Hester's scar is a constant reminder of the brutal crime she witnessed as a child when the famous explorer Thaddeus Valentine murdered her mother in order to steal secrets about how to build a nuclear weapon. This event is the catalyst for Hester's transformation into becoming the ultimate survivor and a woman propelled by the desire to seek vengeance.

Hester is different from the cliché women warriors you see in science fiction and fantasy who tend to be very gorgeous and glamorous. (When you are living in a nuclear wasteland fighting for survival, having that perfect blowout would be the last thing on your mind!) Hester teaches us that our appearance should never hold us back, and not to judge a book by its cover.

When she spills the beans to Tom about her tragic past, Hester reveals that she didn't even cry when she saw her parents being murdered. She has learned to survive by building up so many defenses that she can come across as ruthless or coldhearted to others. By allowing herself to trust others and bring down her walls, she finds friendship and love and is no longer a lone girl against the world.

ABOUT THE AUTHOR

Mortal Engines is the first book in a four-book series written by Philip Reeve, known as the Hungry City Chronicles. Reeve spent more than ten years on *Mortal Engines* before publication in 2001. He originally trained as an illustrator and worked for many years providing illustrations for the popular children's book series *Horrible Histories*.

FACTS

- The name *Mortal Engines* comes from a line in William Shakespeare's *Othello*.
- Peter Jackson purchased the film rights from Philip Reeve in 2001 and quietly worked on the movie for nearly twenty years before it was released in 2018.
- Philip Reeve and his son make a cameo in the film adaptation.

KATNISS EVERDEEN,
THE HUNGER GAMES SERIES

"I am not pretty. I am not beautiful. I am as radiant as the sun."

LESSON LEARNED
Sometimes you've got to fake it till you make it!

PLOT
What was once North America is now a dystopian future with the Capitol Panem maintaining its hold over twelve districts. To maintain its power, each year a boy and girl from each district are selected to compete in a nationally televised event known as the Hunger Games, where each tribute fights to the death until only one remains. For District 12, the poorest of all the districts, Katniss Everdeen volunteers for the seventy-fourth Hunger Games in place of her twelve-year-old sister, Primrose. Katniss must use her resourcefulness and survival instincts to stay alive, all the while plotting to take down the Capitol. Throw in a love triangle and some amazing costume reveals and it's no surprise that The Hunger Games turned into a hit series.

FEMINIST ICON
If you are looking for a strong woman in fiction, look no further. Arrow-slinging, tree-climbing, protect-my-family-at-all-costs, sixteen-year-old Katniss Everdeen does not shy away from a battle to the death and stands as a symbol of resistance against patriarchal systems. It's easy to see why Katniss was quick to get the seal of approval from fans and feminists. She is fiercely independent, scorns feminine frills, and is a master when it comes to a bow and arrow. However, I would argue that what truly makes her a strong woman is not her physical strength, speed, or intelligence, but her ability to communicate, empathize with others, and foster relationships. Katniss's ability to find strength in other women, and to support them in return, makes her a feminist girl on fire.

Family is everything to Katniss. Her sister Primrose (Prim for short) is what gives Katniss the inner strength to fight. Love makes Katniss stronger, not weaker, and not just Peeta's love. It's refreshing to see a young girl whose sole purpose in life is greater than pining over a boy. Peeta needs Katniss just as much as she needs him.

Ultimately, Katniss is a feminist character not because she can put an arrow through an enemy's throat as quickly and cleanly as any man, but because she learns to maintain that strength while opening herself up to the power of mutual support and sisterhood. It's that, perhaps more than anything else, that makes Katniss an ideal role model for girls and an icon for feminist readers.

ABOUT THE AUTHOR

Author Suzanne Collins said she got the inspiration for The Hunger Games book series when she was flipping TV channels between war coverage and reality shows. She adapted the books for film herself and had a say in picking the actors for the movies as well. Even though the protagonist was supposed to be a sixteen-year-old girl, Collins felt that twenty-year-old Jennifer Lawrence was the best fit for the role.

FACTS

- Collins had seriously considered killing off Katniss with those nightlock berries, but after reaching the end of the first book, she realized there had to be a sequel because she knew Katniss would be punished for almost eating the berries.
- The Hunger Games movies were banned in Vietnam by the Vietnamese National Film Board, who likely considered the film's message of political uprising not in line with their government's values.

PART FIVE
School Madams

In every high school across the country, you will find a reading list filled with all the greatest examples of fiction and poetry which supposedly stand any young person in a good stead for life. These lists likely include books like *Of Mice and Men* by John Steinbeck, *Moby Dick* by Herman Melville, *Great Expectations* by Charles Dickens, and a smattering of Shakespeare for good measure.

If we are saying to young people that the only works produced by white men, about white men, have what it takes to be considered classics, what message are we instilling in today's students? In this next section, we explore great feminist classics to discover just how great and ahead of their time their female protagonists truly were.

HESTER PRYNNE,
THE SCARLET LETTER

"She repelled him, by an action marked with natural dignity and force of character, and stepped into the open air, as if by her own free will."

LESSON LEARNED

Act with dignity and tolerance, even when the patriarchy is coming at you from every angle!

PLOT

The Scarlet Letter is set in a village in puritan New England and centers around Hester, a young woman who has borne a child out of wedlock. Hester believes herself a widow, but her husband Roger is very much alive and arrives to find Hester forced to wear the scarlet letter "A" on her dress as punishment for her adultery. She refuses to name her lover to her husband, but he is hell-bent on finding out who Hester's baby daddy is and seeking his revenge. The story follows Hester's path through the years as she faces her punishment and the patriarchy head-on.

FEMINIST ICON

Before there were women fighting against slut shaming, catcalling, and the persecution of women based on their sexual history, there was Hester Prynne, a puritan shunned by her own community after the birth of her illegitimate daughter, Pearl. Though she wasn't the kind of character to lead a revolt or start a women's rights group (this is the seventeenth century, after all), she stood tall and refused to let go of her dignity, despite the nasty whispers and the public shaming she had to endure.

Hester embraced her punishment of wearing the scarlet letter in an act of rebellion and radicalism along with a determination to keep her child! Hester continues to live out her life helping the less fortunate, even though she endures endless taunting and ceaseless ridicule from those in her community. Hester is a true survivor and embodies the feminist ideal that all women should be treated with respect.

ABOUT THE AUTHOR

Nathaniel Hawthorne was born in Salem, Massachusetts, in 1804, and had a very messy family history. His great-great grandfather had been a magistrate during the Salem witch trials of 1692 and had sentenced over a hundred innocent women of being guilty of witchcraft. His published works include novels, short stories, and a biography of his college friend Franklin Pierce, the fourteenth President of the United States.

FACTS

- According to a 1658 law in Plymouth, Massachusetts, puritans really did make people wear letters for adultery.
- *The Scarlet Letter* is packed full of symbolism from the wild woods and the rosebush by the jail to the embroidered "A" itself. It truly is the book that launched a thousand literary essays!
- At the time of publication, people thought the book was scandalous and full of smut. However, this did not affect sales, with the first printing of 2,500 books selling out in just ten days!
- Hawthorne was a very close friend of Herman Melville (author of *Moby Dick*). Although there is no solid proof, many have speculated over the years that the two men were more than just friends.

SCOUT FINCH,
TO KILL A MOCKINGBIRD

"*. . . and by watching her I began to think
there was some skill involved in being a girl.*"

LESSON LEARNED

Fight for truth and justice, even when it gets you in more than a little bit of hot water.

PLOT

Harper Lee's *To Kill a Mockingbird* centers on Atticus Finch's attempts to prove the innocence of Tom Robinson, a black man who has been wrongly accused of raping a white woman in 1930s Alabama. The story is narrated by Scout Finch, a six-year-old tomboy who lives with her lawyer father Atticus and her ten-year-old brother Jem.

FEMINIST ICON

To Kill a Mockingbird often tops the charts of "best books to read before you die." It's won the Pulitzer Prize and is a part of most high school syllabi, all for very good reason. The story stands out because it is narrated by six-year-old Scout, which allows the reader to focus on a young girl's perspective and thoughts. She's not the sidekick, she's the leading role, a refreshing change to other books of the time.

It's also probably not an accident that, although Scout's real name is a very prim and feminine "Jean Louise," everyone refers to her by her nickname. Scout bucks the norm for the era. When other girls pressured her to behave "ladylike," Scout is a strong-headed, plucky tomboy. (As a fellow tomboy myself, I love her!)

By eschewing dresses, walking around barefoot, throwing punches, climbing trees, and being opinionated, Scout serves as an icon of individuality, refusing to follow restrictive traditions just because of her gender. When her Uncle Jack, upset that she's asking too many questions, asks her if she wants to grow up to be a lady, Scout nonchalantly replies, "Not particularly." However, after watching Calpurnia (the Finch family's housekeeper), Scout realizes that being a girl actually involves having positive traits rather than just being the lesser of the two sexes.

Growing up in Maycomb, Alabama, during a time of extreme racism, Scout defends the weak and fights for the innocent. She searches for the goodness in other humans even when society tells her not to look beyond skin color.

ABOUT THE AUTHOR

Harper Lee was born and raised in the small town of Monroeville, Alabama, where her father was an attorney. She based *To Kill a Mockingbird* on the people and places she experienced during her childhood. In fact, the plot of *To Kill a Mockingbird* is based on one of her father's cases involving two black men convicted of murder.

FACTS

- *To Kill a Mockingbird* is one of the bestselling books of all time, with more than forty million copies sold.
- It was long thought that *To Kill a Mockingbird* would be Lee's first and last novel but, at eighty-nine years old, HarperCollins published her rediscovered manuscript *Go Set a Watchman* in 2015 as something of a sequel to *Mockingbird*. It was an instant and controversial bestseller.
- Harper Lee had a long friendship with writer Truman Capote, author of *Breakfast at Tiffany's*. They were childhood friends. She modeled the neighborhood boy Dill in *To Kill a Mockingbird* after Capote.

BECKY SHARP,
VANITY FAIR

"A person can't help their birth."

LESSON LEARNED

It's better to go after life full throttle than to sit quietly waiting for something to happen.

PLOT

An orphan of low birth, Becky Sharp is determined to raise herself to the top of high society to gain wealth and riches.

FEMINIST ICON

At a time when women are pawn pieces in a patriarchal society's game of chess, Becky Sharp throws the board across the room. The star of *Vanity Fair* is one of the most repugnant characters in literature, yet I adored her from the glorious moment she tosses her dictionary— a parting gift—out of the carriage window as she exits Miss Pinkerton's Academy for Young Ladies. Becky Sharp is everything a Regency woman ought not to be: selfish, egotistical, ruthless, ambitious, and driven only by money. Her last name is no coincidence.

While the happiness of most leading ladies rests on their making a love match, Becky is unsentimental about matters of the heart. Men are mere steppingstones in her quest for social status and material gain. She is also painted as unfeminine and unmotherly. For these reasons, she's called an antiheroine. However, this does her a massive disservice. Any man with her traits would be lauded as a hero. It's just that Becky is that most challenging of things: a protagonist who is unscrupulous, self-interested, and female. She makes for a literary breath of fresh air who, two hundred years later, is just as relevant as she ever was. She is shameless, delightfully morally bankrupt, and we love her for it.

ABOUT THE AUTHOR

William Makepeace Thackeray was born in 1811 in Calcutta, India, where his parents made their fortune in the East India Company. After the death of his father when he was five, he was sent to England to live with his aunt. While highly educated, attending the prestigious Cambridge University, Mr. Thackeray never completed his degree before starting his early career in journalism.

Thackeray took the name *Vanity Fair* from another story: John Bunyan's *A Pilgrim's Progress*. Bunyan's story is a Christian allegory of a man attempting to reach Heaven but being tempted and tried and generally messed with en route. One of the "towns" he stops in is Vanity Fair, where the Devil sells things to tempt human desires. It's meant to be a representation of all our attachments to worldly things and vanities in one place, hence the name.

FACTS

- Thackeray's greatest work has lent its name to the popular cultural and fashion magazine *Vanity Fair*.
- Oscar Wilde himself speculated on which real-life female Becky Sharp was based on.
- Mr. Thackeray was a gambler and is rumored to have lost some of his inheritance to the habit.
- Thackeray was also an illustrator and a student of law and fine arts.

CELIE,
THE COLOR PURPLE

*"I'm pore, I'm black, I may be ugly and can't cook,
a voice say to everything listening. But I'm here."*

LESSON LEARNED

Although the world is full of suffering, it is also full of people overcoming it.

PLOT

The Color Purple documents the traumas and gradual triumph of Celie, an African American teenager raised in rural isolation in Georgia. The novel weaves an intricate mosaic of women joined by their love for each other, the men who abuse them, and the children they care for.

FEMINIST ICON

The Color Purple is filled with themes of racism, sexism, and violence against women. Among all these horrors, Celie shines out like a bright, feminist star, drawing us all in. Hers is a story of pain and perseverance in the face of extreme abuse, oppression, loneliness, and pain. Celie is the living embodiment of what it means to be a strong woman and teaches us that even in the darkest of days there is still hope for a better tomorrow.

Celie narrates her life through painfully honest letters to God, allowing us to experience firsthand the raw emotion and pain she is going through. Celie's life has been far from easy. She was raped as a child by her father, has her two children taken from her, and is married off to a stranger who abuses her constantly.

Celie is not alone in experiencing this cycle of abuse. All the women in *The Color Purple* suffer from living under the patriarchal rules enforced by society. These women are not regarded as free and have no control over their own lives. Celie believes that the only way to survive in her husband's house is to obey his rules, tolerate the beatings, and remain silent. However, as she speaks to other women, she finds her voice and starts to stand up for her rights. Her relationship with the other women in the novel, including Nettie (her sister), Shug, and Sofia is a beautiful and moving example of the power of sisterhood and how, if we build each other up instead of tearing each other down, we can help one another thrive.

Celie embodies the strength of the human spirit and the power of forgiveness. We see her transform from a wounded, beaten woman to a strong, independent, and loving individual.

ABOUT THE AUTHOR

Poet, essayist, and novelist Alice Walker was born in 1944 in Eatonton, Georgia. The youngest daughter of sharecroppers, she grew up poor, with her mother working as a maid to help support their family's eight children. Living in the racially divided South, Walker showcased a bright mind at her segregated schools, graduating from high school as class valedictorian. Walker is a feminist and vocal advocate for human rights, and she has earned critical and popular acclaim as a major American novelist and intellectual. Her other works such as *In Search of Our Mothers' Gardens* and *You Can't Keep a Good Woman Down* showcase Walker's own brand of feminism called "womanism," which advocates for the rights of women of color.

FACTS

- *The Color Purple* won the Pulitzer Prize in 1982, although the book has been banned in schools and libraries across the United States due to its graphic sexual content and situations of violence and abuse.
- Alice Walker and her husband Melvyn Rosenman Leventhal, a Jewish civil rights attorney, had the first legal interracial marriage in Mississippi in 1967.
- In 1985, Steven Spielberg adapted *The Color Purple* into a movie, which became a box office hit grossing $142 million. In 2005, the novel was turned into an award-winning Broadway musical that ran for three years.
- In 2018, Spielberg, Oprah Winfrey, Quincy Jones, and Scott Sanders announced that they would be transforming the Broadway musical into a big screen musical.

ELIZABETH BENNET,
PRIDE AND PREJUDICE

"My courage always rises at every attempt to intimidate me."

LESSON LEARNED
Hold fast to your own values and don't let anyone dictate your decisions.

PLOT
Pride and Prejudice is a romantic novel that follows the young Elizabeth Bennet. She is one of five unmarried daughters who, much to their mother's distress, are set to become destitute upon their father's death unless they can make an advantageous marriage. The novel revolves around the importance of marrying for love instead of money and Elizabeth's refusal to conform in society.

FEMINIST ICON
Pride and Prejudice's Elizabeth Bennet is smart and imaginative, and despite the best efforts of those around her, she refuses to give in to the expected role of a woman. She's opinionated and outspoken, and her boldness became an early example of sticking it to the patriarchy!

Let's be honest, when people say they would like to be Elizabeth Bennet, they're often more interested in Mr. Darcy (a.k.a. Colin Firth in that shirt!). At a time when women were forced to marry for financial stability rather than love, Lizzie chooses the more difficult, uncertain option—and proves the doubters wrong. She champions the downtrodden and she fights hard for her own happiness. Neither the temptations of wealth and prestige nor the intimidation of scary aunts like Lady Catherine de Bourgh can shake her convictions. She controls her own narrative and maintains her identity in marriage.

ABOUT THE AUTHOR
Jane Austen was born into the Georgian era in 1775. She was one of eight children in a very close-knit family. She wrote six novels in total and developed a substantial fan base during her life. *Emma*, her fourth novel, was even dedicated to the Prince regent, an admirer of her work. Although Jane Austen never married, she was no stranger to romance. She had several romantic interests over the years, one of note being Harris Bigg-Wither who proposed to Austen. However, after initially accepting his offer of marriage, she changed her mind overnight, later telling her

niece that she found him "very plain in person—awkward, and even uncouth in manner."

FACTS

- Jane Austen's novels were originally published anonymously. Women were not welcome in the world of writing, and it was considered improper for those who did. Thankfully, the pen proved mightier than the penis and Jane is accredited as one of the greatest writers in HERstory.
- Jane Austen brewed her own beer. Her specialty was a "spruce beer" made with molasses, giving the beer a sweet taste. Cheers to that!
- *Pride and Prejudice* was originally titled "First Impressions."
- Although *Pride and Prejudice* was a massive hit upon publication and has been in continuous print ever since, it did not make Austen much money, as she sold the book outright to her publisher rather than taking royalties.

WILHELMINA (MINA) HARKER,
DRACULA

"Ah, that wonderful Madam Mina! She has man's brain—a brain that a man should have were he much gifted—and a woman's heart."

LESSON LEARNED
Do not let your fears control you.

PLOT
The Gothic novel *Dracula* by Bram Stoker was first published in 1897 and became the basis of an entire genre of literature, television, and film. The novel is written as a series of journal entries, letters, and telegrams by the main characters. It begins with Jonathan Harker, a young English lawyer, as he travels to Transylvania. Harker plans to meet with Count Dracula, a client of his firm, in order to finalize a property transaction. However, things go downhill once Harker realizes he is a prisoner in Dracula's castle. Fearing for his life, Harker escapes only to find out later that Dracula is moving to England where he plans on turning Harker's wife, Mina, into a vampire.

FEMINIST ICON
By today's standards, Mina Harker might not appear to be as strong as some others featured in this section, but for her time in repressive Victorian society where the class system ruled, I would argue that she is an incredible source of feminine strength and intelligence. She is the first to figure out Dracula's plans long before the men in the story. (If only the men had listened to her!)

In one of the most backhanded compliments, the ruthless vampire killer Van Helsing himself says, "She has a man's brain—a brain that a man should have were he much gifted—and a woman's heart." And in truth, without Mina's help, it is clear than none of the men would have been able to find and kill Dracula.

ABOUT THE AUTHOR
Bram Stoker was born in Dublin, Ireland, and for an early part in his career worked in Dublin Castle. His gothic masterpiece *Dracula* has gone on to define a genre of horror with one of literature's most iconic characters: a blood-slurping, shapeshifting, garlic-hating vampire. Without this work, there might never have been any *Vampire Diaries*, *A Discovery of Witches*, or *Twilight*.

FACTS

- The inspiration for Dracula came to Bram Stoker after a suspected bout of food poisoning, which led to some nightmarish dreams.
- Vampires and Frankenstein may come from the same exotic Victorian mini break. In 1816, on a gloomy day in Lake Geneva, Lord Byron proposed a ghost story contest that led to Mary Shelley writing *Frankenstein*. It was also the birth of *The Vampyre* by John Polidori, his first-ever vampire story written in English.
- Dracula has appeared in just short of three hundred films, making him the most portrayed character in film, just ahead of iconic detective Sherlock Holmes.

PART SIX
Fierce Females of Fantasy

Kings, princes, wizards, and warlocks underpin the fantasy fiction genre. The only problem with this is that it leaves very little room for the women!

Fantasy fiction is centered on a hero. The hero must be brave, strong, never mean or malicious, and possessed with a chivalrous code of honor that requires them to come to the aid of the weak and oppressed when nobody else will. These characteristics coincide with conventional masculine values in which men are expected to throw aside their emotions and fight at all costs. This warped way of defining a hero has led male characters to dominate in heroic fiction from the earliest of times, while female characters are often limited to roles that are defined in relation to the male hero, from the sexy temptress and the damsel in distress to the virginal bride.

Fantasy fiction has long been calling out for some powerful women to reclaim their stolen kingdoms, combat misogynistic monarchies, and forge their own destinies. While the genre has predominately seen male authors dominate the charts with male-centric plots, there are a few breakaway characters that really shine in the realm of fantasy fiction.

ÉOWYN,
THE LORD OF THE RINGS FRANCHISE

"But no living man am I! You look upon a woman."

LESSON LEARNED

If you want a job done well (or a dark sorcerer murdered), do it yourself!

PLOT

The future of a civilization rests in the fate of the One Ring, which has been lost for centuries. Powerful forces are unwavering in their search for it, but fate has placed it in the hands of a young Hobbit named Frodo Baggins. Frodo, along with his friends, must take on the daunting task of destroying the One Ring in the fires of Mount Doom.

FEMINIST ICON

Most of the women in LOTR are nothing more than background characters, but there are three (four, if you count Shelob, but I don't because she's a spider) who have significant roles to play in the plot. But to paraphrase Tolkien himself, if there can only be one woman to rule them all, for me it has to be Éowyn. She's brave, she's rebellious, and most importantly of all, she's gender nonconformist! Tolkien might not have had much time for women as some have claimed, but in Éowyn he showed that he certainly had some understanding of the frustrations women experience when they are expected to conform to sexist stereotypes. Desperate to be allowed to fight alongside the men and thwarted from doing so by both Théoden and Aragorn, Éowyn's reluctance to bow to patriarchal conditioning ultimately saves the day.

When Aragorn asks Éowyn what she fears if it's not pain or death, she replies: "A cage. To stay behind bars." It is this fierce spirit that leads Éowyn to disguise herself as a man and go to battle. And it's a good job she does because when she's face to face with the Witch-King in the Battle of the Pelennor Fields, it's clear that the war couldn't have been won without her. As Éowyn draws her sword, the Witch-King screams "Hinder me? Thou fool. No living man may hinder me!" But Éowyn simply laughs at him, and retorts, "But no living man am I! You look upon a woman." And with that she pulls off her disguise, whips her hair (picture one of those shampoo ads) and plunges her sword into his heart, killing him.

ABOUT THE AUTHOR

John Ronald Reuel Tolkien was born in 1892 in South Africa before moving to England at the age of three with his family. However, his childhood was not a happy one, as he was orphaned as a young boy. Tolkien was sixteen when he met his lifelong partner Edith, who his guardian initially forbade him from dating. The pair were twenty-one when they finally got together, although Edith was at that time engaged to someone else. The pair's lifelong romance is immortalized in Tolkien's other great work *The Silmarillion*. The names of the two main characters (Beren and Lúthien) even feature on Tolkien's and Edith's gravestones.

FACTS

- The Nazis would only publish *The Hobbit* in German if Tolkien could prove that his lineage was pure. Tolkien replied with the F.U. of all letters, writing that he regretted he didn't have more Jewish ancestry and calling Hitler a "ruddy little ignoramus."
- Tolkien was quite the linguist. He invented fourteen languages of his own such as the Elvish languages Quenya and Sindarin, which he used extensively in his writing. He could also read and write fluently in over twenty languages.
- While *The Hobbit* was a children's book, The Lord of the Rings trilogy was a lot darker and meant for adults.
- Tolkien was very close friends with author C. S. Lewis.
- Tolkien was a cadet and helped with the parade route of King George V's coronation in 1910.
- *The Lord of the Rings* is one of the bestselling books of all time, with over 150 million copies sold.

ARYA STARK,
A SONG OF ICE AND FIRE SERIES

*"I am a wolf and will not be afraid . . .
But I'm not a lady. I never have been. That's not me."*

LESSON LEARNED

Experiences of trauma can lead to resilience and strength.

PLOT

If I had to sum up this epic series of novels in three words, they would have to be murder, sex, and dragons! George R. R. Martin's A Song of Fire and Ice series takes place on the fictional continents of Westeros and Essos with each chapter being told from the perspective of one of the characters. There are three main stories that weave into an epic tale: a dynastic war among several families for control of Westeros, the rising threat of the supernatural Others in northernmost Westeros, and the ambition of Daenerys Targaryen, the deposed king's exiled daughter, to assume the Iron Throne.

FEMINIST ICON

With Lyanna Mormont, Sansa Stark, Daenerys Targaryen, Brienne of Tarth, Lady Stoneheart, Cersei Lannister, and Ygritte, there are several strong female characters in George R. R. Martin's A Song of Fire and Ice series who are all inspirational in their own ways, but fan favorite Arya Stark is arguably the strongest. She is dominant, powerful, and in charge of her own fate. She's violent and vulnerable, charming and outrageous, independent and bound to the family she loves. It is these characteristics that define the flawed Arya, who fights tooth and nail to survive.

When we first meet Arya, she is a child who refuses to conform to the gender roles of sexist society. Forced to wear a dress and practice needlework with her sister, Arya knows that is not her destiny. She instead excels at tasks that girls aren't supposed to do, such as being able to shoot a bow and arrow better than her younger brothers. When her older "brother" Jon Snow gifts her a sword, she names it Needle in a jab to the ridiculous gender constructs that as a girl she should be more concerned with needlework than sword play. After witnessing her father's beheading, she dedicates her life to seeking revenge. Arya's drive to enforce justice helps her channel her anger toward her goals and is a reminder that if you desire something deeply, nothing can stop you. Whether it is overcoming fear, excelling at a new talent, or simply becoming the ultimate assassin to dispatch all those who wronged you, there is a lot to learn from this fierce female!

ABOUT THE AUTHOR

George R. R. Martin is often referred to as "the American Tolkien." More than ten years before A Song of Ice and Fire debuted in 1996, Martin wrote a book called *The Armageddon Rag* in 1983. Though it was a critical disappointment, producer Phil DeGuere was interested in adapting the project for film with Martin's help. While that never came to fruition, DeGuere thought of Martin when they were rebooting *The Twilight Zone* in the 1980s and brought him on board to write a handful of episodes.

FACTS

- Martin's inspiration for the series included *The Wars of the Roses* and the series of French historical novels The Accursed Kings by Maurice Druon.
- A Song of Ice and Fire sold over ninety million copies world-wide and been translated into forty-seven languages.
- As a child, Martin wrote and sold stories to other neighbor-hood children. He initially charged a penny but later (after popular demand) raised his prices to a nickel.
- In the twenty-three years since he started writing the series, Martin has written an estimated 1.8 million words of A Song of Ice and Fire.
- A Song of Ice and Fire almost had no dragons! Martin says he debated not including them but was convinced by his friend and fellow author Phyllis Eisenstein.
- The next installment of the series is called *The Winds of Winter*, although no release date has yet been given.

LYRA BELACQUA,
HIS DARK MATERIALS

"Good and evil are names for what people do, not for who they are."

LESSON LEARNED

Don't just follow the rules but think for yourself.

PLOT

Lyra Belacqua is the heroine of Philip Pullman's trilogy His Dark Materials, an epic story set across parallel universes. In the world in which Lyra grows up, there is a magical force known as "dust" that enables people to be bound to a dæmon (an external physical manifestation of a person's inner self). Lyra and her dæmon Pantalaimon (known as "Pan") find themselves at the heart of a cosmic war between Lord Asriel (her distant father) on one side and a deity figure known as "The Authority," a ruling religious order, on the other.

FEMINIST ICON

There are some massive philosophical quandaries at play in His Dark Materials. To this day, I am not sure I fully understand all the symbolism and metaphors regarding religion, power, and loss that feature across the series. However, what I am sure about is what an amazing character Lyra is!

Lyra inhabits a role often reserved for male characters: the selfless, brave adventurer determined to defeat evil. She rejects traditional femininity, repeatedly ruining her dresses when roaming the rooftops of Jordan College in Oxford where she has grown up essentially an orphan. However, it is unsurprising that growing up in this solid patriarchal structure of Jordan College has left Lyra with internalized misogyny. She is at first dismissive of the female scholars she encounters and is mistrusting of most women, but Lyra learns and grows in so many ways across the series, examining many aspects of what she has always believed along the way.

Lyra is portrayed as a tomboy, returning from every escapade with dirty fingernails and tangled hair, rejecting the gendered constraints that are in force around her. Lyra marks out a space for herself in the male world of knowledge, power, and adventure and rejects gendered constraints upon her body and mind. Most of Lyra's actions are rebelliously deviant and push the boundaries of societally sanctioned female behavior. From fighting with the Gyptians, rescuing children from the Gobblers, freeing the souls of the dead, and riding armored bears, Lyra repeatedly spits in the face of patriarchal expectation.

One of the most symbolic metaphors that can be drawn from Lyra is how she brings redemption to all of womankind in the rewriting of the story of Eve and the fruit from the tree of knowledge in Genesis. Reflecting Pullman's own views on religion and free will, rather than Eve being the cause of all pain and suffering in the world when she took a bite of the forbidden fruit, as Christian doctrine would have you believe, Eve was in fact a heroine who freed humanity from the restrictive constraints of the Garden of Eden, allowing humanity's intellectual growth. Lyra embodies the "second Eve" through her use of the alethiometer, which allows her to seek the truth.

It's easy to feel so connected to the magical world Lyra inhabits because we see her innermost thoughts through the conversations she has with Pan. While Lyra is hotheaded and impulsive, Pan is cautious and levelheaded. It is these contradictory personality traits we all have inside ourselves that dictate our behavior. I have often imagined talking to my own inner dæmon while trying to stay calm and not overreact to situations. But then again, there is a lot of fun to be had in taking after Lyra's example and calling people out for what they really are.

ABOUT THE AUTHOR

British author Philip Pullman is a heavyweight in the realms of fantasy fiction. He lived in many different places during his childhood, including Zimbabwe and Australia, before his family finally settled in North Wales. He read English at Exeter College, Oxford (thinly disguised as Jordan College in *His Dark Materials*). Philip is a disciplined writer and will not leave his desk until he has written at least three pages of A4 (old-school pen and paper). At the end of a working day, he will always make sure to write at least one line on a new sheet of paper so that he is not faced with a blank sheet of paper when he starts work the next day.

FACTS

- *His Dark Materials* is a retelling of Milton's epic poem from the seventeenth century telling of *Paradise Lost*, the story of Adam and Eve, and of Satan's banishment from heaven. Pullman read the book as a teenager and fell in love with it. Years later, he got the idea to write a story that flipped the poem on its head (and added a few warrior polar bears along the way!).

- Although *His Dark Materials* has been marketed as young adult fiction, and the central characters are children, Pullman wrote with no target audience in mind.

PART SEVEN
Comedy Queens

We can decorate our homes with inspirational quote posters and accessories in homage to feminism, but there are times in life when you just want to curl up on the sofa and watch something uplifting and funny. This section is dedicated to the magnificent comedy queens of the big and small screen whose feminist feel-good factor is no laughing matter.

Nowadays there are so many good examples of incredible female-led comedy shows and movies that not only smash the Bechdel test, but also challenge the preconceptions of how female characters "should" behave.

Like many, I was raised in a society where women have been repeatedly told the dreaded "women aren't as funny" argument. This, of course, is a big steaming pile of bs. Comedians like Tina Fey, the former head writer of *Saturday Night Live* who created *Mean Girls* and the sitcom *30 Rock*, is one of the leading voices in the fight for feminism in comedy. When interviewed by *Town & Country* magazine about her experience as a woman in comedy, Tina Fey replied, "People really wanted [Amy Poehler and I] to be openly grateful—'Thank you so much!' And we were like, 'No, it's a terrible time. If you were to really look at it, the boys are still getting more money for a lot of garbage, while the ladies are hustling and doing amazing work for less.'"

Comedy has changed over the years. The repertoire of women isn't limited to self-loathing or man-hating anymore. The humor is more eclectic, serene, and organic. As comedy has opened up, women who once might not have dared write comedy or writers who hadn't considered performing are emboldened to take the stage. As a result, we are seeing more and more of ourselves reflected in popular culture.

ELLE WOODS,

LEGALLY BLONDE

"What, like it's hard?"

LESSON LEARNED

Having an interest in frivolous things doesn't make you stupid.

PLOT

Legally Blonde burst onto our screens in 2001 and has since become a cult movie for a generation. The movie starts with Elle Woods, a sorority sister who thinks she is just on the cusp of getting her happily ever after with her college boyfriend Warner. However, instead of proposing, he breaks up with her because he doesn't think Elle is "serious" enough for him as he starts Harvard Law School. Desperate to win Warner back, Elle studies for and passes the LSAT, applies to Harvard, and is accepted. However, while studying she learns more than just the law and sets her sights higher.

FEMINIST ICON

Legally Blonde is the feminist fairy tale all young girls should grow up watching. Although Elle's journey has a rocky start, she is the epitome of the underdog, proving that hard work and resilience is what drives you to succeed—not the opinions of others. There are so many life lessons to learn from *Legally Blonde* that it's hard to narrow it down to just a few but here goes:

First and foremost is that Elle is never scared to be herself. She is unapologetic for her can-do positive attitude and bubblegum pink persona (it is her signature color, after all). Despite being ridiculed at Harvard by classmates and faculty alike for her feminine taste and love of pink, Elle never allows others to impact her self-worth. She is fearlessly true to herself and proves time and time again that being stereotypically "girly" by no means undermines her capability. In order to succeed in most workplaces, women are often expected to present a more masculine persona. Elle Woods rejects this expectation and uses her femininity to her advantage in a patriarchal setting.

Elle also teaches us that having interests in subjects such as fashion, beauty, and celebrity gossip by no means detracts from our intelligence. Elle is not only smart enough to understand the nuances of complex law but she also knows what dresses were so last season; proving that intellect and frivolity are not mutually exclusive. Her intelligence is constantly questioned by those around her, from her ex-boyfriend and his new girlfriend to a sales assistant who assumes she's stupid because of how she looks, but Elle is stronger and smarter than anyone realizes. She was not only the president of her sorority, showing great leadership skills, but she also had a 4.0 GPA and scored a 179 on her LSAT (out of 180 possible points), making her a top candidate with tremendous work ethic.

The film was also ahead of its time with the relationship between Elle and Warner's new preppy fiancée Vivian. Vivian and Elle were set up to not only compete over Warner, but also grades and career opportunities. However, unlike the cliché archetypes of women tearing each other down, the two become united after witnessing the blatant sexism in their field. When Elle's professor is sexually inappropriate with her in his office, suggesting that she needs to cross professional boundaries if she wants to make it as a lawyer, she slaps away his hand and immediately reproaches him for being a "pathetic asshole." Elle considers leaving Harvard after this encounter, but she is ultimately reassured by her female professor to keep on fighting.

One of Elle's most striking qualities is her optimism. She sees the best in people, despite others sometimes refusing to see the best in her. Without optimism, feminists today could not have made such significant progress. Elle Woods inspires women of all ages to strive for what they want, and although Elle's initial goals are perhaps less feminist-friendly, what she learns on her journey is of huge significance. She realizes the importance of creating her own success instead of seeking validation from others. And most importantly she learns to love herself first, recognizing that she doesn't need a man's approval to be happy.

FACTS

- The film was based on a book written by Amanda Brown that was inspired by her own experience of attending Stanford Law School. True to form, she wrote the manuscript on pink paper with her pink fluffy pen.
- During Elle's Harvard Law School admission video, Elle's sorority Delta Nu votes on opposing the change from Charmin toilet paper to a generic brand. That scene was based on cowriter

Karen McCullah's experience in a sorority at James Madison University.

- Reese Witherspoon (a feminist icon in her own right) has talked about how "at least once a week I have a woman come up to me and say, "I went to law school because of *Legally Blonde*."

- Witherspoon filmed the movie shortly after having her first child Ava and was very much sleep deprived throughout filming.

- According to the writers, the "bend and snap" was a spur-of-the-moment, drunken creation.

- *Legally Blonde 3* is in the works and is being written by the amazingly talented Mindy Kaling.

MINDY LAHIRI,
THE MINDY PROJECT

*"No man tells me what to do with my body,
only women's magazines can do that."*

LESSON LEARNED

When life gets you down, lying on the floor is a valid coping mechanism.

PLOT

The Mindy Project is a rom-com bursting with feel-good moments, fashion inspiration, and a renewed sense of feminism for all the hopeless romantics out there. Set in New York City, the show follows the hectic life of Dr. Mindy Lahiri as she tackles her career in a small medical practice as an OB/GYN surrounded by quirky colleagues. Mindy, who had grown up devouring any rom-com featuring a female lead on the hunt for Mr. Right (much like the show's creator and star Mindy Kaling), takes her own hunt to find Mr. Right into the twenty-first century. Just because she wants her life to be a *You've Got Mail* sequel doesn't mean she has to compromise on her values and play second fiddle to a Mr. Right.

FEMINIST ICON

Mindy is not perfect and that is why I feel so drawn to her as a character. She is vain, materialistic, and at times downright ruthless. Mindy is prone to the occasional floor meltdown, inhaling doughnuts, and being overly dramatic about any situation (and let's be honest, we have all been there!). These flaws along with her complex and well-rounded personality make her such a strong woman that you can't help but fall in love with her. *The Mindy Project* treats audiences to a different kind of rom-com where, instead of our plucky heroine finding completeness in someone else, she learns to find it within herself.

As an OB/GYN, Mindy is a great advocate for women's health issues. Many episodes tackle important issues such as women's fertility, the stigma of breastfeeding in public, ownership over one's body, a woman's right to choose, and access to affordable health care. As well as exploring the sexism which exists in medicine, *The Mindy Project* also tackles the issue of race. In one episode (*Mindy Lahiri Is a White Man*), Mindy is passed over for a promotion and in a Freaky Friday situation goes to bed wishing things were different, only to wake up as a handsome white man. Mindy quickly discovers the perks that come with white male privilege. The episode was funny, ridiculous, and absurd while also pivotal, telling, and groundbreaking, capturing what many women of color experience daily. It is no surprise that such a feminist icon was created and played by the talented actress and screenwriter Mindy Kaling. Throughout her career, Kaling has been vocal about feminism and women's rights. In a 2016 Watermark Conference for Women, Mindy delivered a keynote speech where she spoke about her own opinions on what it means to be a strong woman: "A lot of people sometimes think that you can't be interested in things like fashion or things that are traditionally feminine and also be a strong woman . . . because that means you are doing that for a man . . . I wish that we didn't inextricably link being interested in those things and being a strong woman." This is one of my favorite things about watching *The Mindy Project*: I could not get enough of her wardrobe! Throughout the show's six seasons, Mindy's outfits never failed to amaze. She was not wearing them to attract the male gaze but her clothes were rather an extension of who Mindy was inside: bold, vibrant, and confident. In fact, Mindy's relationship with her body features heavily throughout the series in one of the most realistic portrayals in television. Although Mindy does find faults and criticizes her own appearance time to time, she is all about celebrating her curves and her skin tone.

FACTS

- When Kaling originally joined *The Office,* she was the only woman working on the writers' staff. Greg Daniels, the show-runner, has said that she was the best writer on the show.
- The surname Lahiri comes from Kaling's favorite author, Jhumpa Lahiri. She reportedly reads all of Lahiri's books the week they come out.
- Kaling's mother was an OB/GYN and was the inspiration behind Kaling's character in *The Mindy Project*.
- Mindy Kaling has also gone on to cocreate *Never Have I Ever* alongside Lang Fisher, which is loosely inspired by her own Indian American childhood. The show is another feminist coming-of-age hit which centers around Devi (played by Maitreyi Ramakrishnan).

LESLIE KNOPE,
PARKS AND RECREATION

"I am a goddess. A glorious female warrior. Queen of all that I survey.
Enemies of fairness and equality, hear my womanly roar."

LESSON LEARNED
Playing it cool is underrated. Know what you want and go at it full force.

PLOT
Parks and Recreation follows the glorious Leslie Knope (played by the fantastic feminist Amy Poehler) who is the deputy director of the Parks and Recreation department of the small town of Pawnee, Indiana. She puts most of her energy into helping the at-times-ungrateful residents of Pawnee while pursuing her dream to make a difference in politics. *Parks and Recreation* is a feel-good comedy with real heart and a wonderful cast of strong women.

FEMINIST ICON
Once on a group project at university, a guy who was annoyed at the level of excitement I was bringing called me Leslie Knope. It was intended as an insult, but to this day I count it as one of the best compliments I have ever received. It baffles me how anyone could think calling someone Leslie was an insult! Leslie is likable, enthusiastic, and hugely passionate about her town and the people who live there. She is an emblem of feminism that stands up for her beliefs, puts her friends first, and has a full-blown love affair with waffles! If I could manifest to be even 10 percent of the awesomeness that is Leslie, I would be winning in life.

Throughout the seasons, Leslie Knope has been a fantastic fictional feminist on the small screen, showing us that women can be career-focused, smart, and in charge as well as loving life and men along the way. In a culture where reality stars and models are heralded as living idols, Leslie proudly displays photos of her own feminist icons on her wall of "inspirational women" which features Eleanor Roosevelt, Nancy Pelosi, Ruth Bader Ginsburg, Michelle Obama, Nancy Pelosi, Condoleezza Rice, and herself, as Leslie is big enough to admit that she is an inspiration to herself. That energy is something we can all do with a sprinkle of. Leslie's self-confidence is one of her most positive and inspirational qualities. When she wants something, she goes after it, and although she may experience some setbacks along the way, she always ends up getting there in the end. Her self-assurance is inspiring.

In one episode when her colleague and friend Ron Swanson is in charge of his own boys-only wilderness troop, the Pawnee Rangers, Leslie sees that girls were denied the chance to join, so she starts her own Girl Scouts inspired group, the Pawnee Goddesses. Leslie shows the world that to declare yourself a feminist is not a dirty word and that feminists aren't all angry, ball-busting, hairy-legged man-haters.

In another stroke of genius, sick of the emphasis on finding a special someone to celebrate Valentine's Day with, Leslie invents the now well-established holiday Galentine's Day on February 13. A perfect day to celebrate female friendships!

Leslie may be happily married to Ben, but the strongest relationship on the show is her friendship with Ann (Rashida Jones). Leslie and Ann support each other in everything they do, and never fail to show their deep love and appreciation for each other. Leslie is possibly the world's greatest friend (she even has a calendar of obscure friend anniversaries), and despite her marriage and time-consuming job, her friends always come first. As Leslie would put it, "uteruses before duderuses, ovaries before brovaries."

FACTS

- *Parks and Recreation* was written and created by Greg Daniels and Michael Shur. Daniels also notably brought *The Office* to life on NBC.
- The role of April was written specifically for Aubrey Plaza after one of the casting directors called Schur saying, "I just met the weirdest girl I've ever met in my life. You have to meet her and put her on your show."
- *Moxie* was produced, directed, and stars Amy Poehler and is based on the 2015 book by Jennifer Mathieu of the same name.
- Amy Poehler is the cofounder of Amy Poehler's Smart Girls, an organization and website aimed at helping young women with life's problems in a funny but informative way. Poehler says the idea came out of wishing we had a time machine to go back and tell our younger versions of ourselves it was going to be okay.

- Multiple versions of the scene where Leslie meets VP Joe Biden in "Leslie vs. April" were shot just in case Obama and Biden lost the 2012 election.

PHOEBE BUFFAY,

FRIENDS

"I wish I could, but I don't want to."

LESSON LEARNED
Honesty is the best policy, not only with others but also yourself.

PLOT
Friends was the TV show that defined a generation, made coffee shops the most wanted hangout, and launched an iconic haircut, "The Rachel." Based on six twenty-to-thirty-something-year-old friends living in Manhattan, the show followed these characters through just about every life experience imaginable: love, marriage, divorce, children, heartbreaks, fights, new jobs, and loss.

FEMINIST ICON
Phoebe may be packaged as the quirky, weird, and wonderful friend, but she is more than meets the eye. Phoebe had an unconventional upbringing. Her biological mother gave her and her twin sister up to her own best friend, her biological father abandoned them all, her adoptive mother killed herself when she was thirteen, and her stepdad went to prison, leaving poor little Phoebe to fend for herself on the mean streets of New York City. Phoebe is a survivor and does not let any of her trauma deflect from her positive attitude. Phoebe is never ashamed of her past. In fact, she talks about it often throughout the show. She didn't grow up as lucky as the others, but that doesn't faze her one bit. This admirable characteristic shows that there is absolutely nothing wrong with being proud of where you come from, whether that's a perfect white picket-fence house in the Hamptons or a cardboard box in Queens.

In all the 236 episodes of *Friends*, I can't think of a single one in which Phoebe wasn't 100 percent honest or didn't say exactly what was on her mind. Whether it was voicing an unpopular opinion or as simple as expressing her desire *not* to do something, Phoebe never beats around the bush and has mastered the art of saying no.

When she finally found "the one" in Mike, she was strong enough to say goodbye to him when he confessed that he never wanted to get married. She stays true to her own dreams and the future she sees for herself even though it leads to heartbreak. However, Phoebe and Mike do get back together after Mike changes his mind, but their relationship wouldn't have felt as defined if Phoebe had just succumbed to whatever Mike wanted (as Monica and Rachel did time and again with their significant others).

Phoebe's unflappable confidence is one of her most admirable traits. The amount of time we all waste agonizing over decisions or stressing out about awkward encounters accomplishes nothing but sleepless nights, so be more like Pheebs and don't fret over the little things (even if it hasn't been your day, your week, your month or even your year!).

FACTS

- Lisa Kudrow, who plays Phoebe Buffay, had to learn how to play the guitar for the role. And thank goodness she did, as who would want to be in a world without "Smelly Cat"?
- The cast of *Friends* was the first to negotiate pay as a whole group. By the final season, each cast member was earning $1 million per episode. Some much needed pay equality in Hollywood!
- Phoebe Buffay had a twin sister called Ursula. Kudrow was already playing Ursula in the show *Mad About You* when she was cast in *Friends* and was encouraged by NBC to play both roles.
- Kudrow got pregnant with her son, Julian Murray, in 1997. Kudrow was dubious about Phoebe getting pregnant, too, but the writers decided to have Phoebe act as a surrogate for her brother's triplets.
- There are several replicas of Central Perk café around the globe, from New York to Beijing.

PART EIGHT
Badass Babes

Back in the mid-fifties when the word "badass" originated, it was not something you wanted to be. The term referred to a bully or dangerous individual and was only used to describe men. Seventy years on, however, "badass" is no longer a put-down for men but rather a battle cry for the fierce feminists who put themselves first rather than appeasing others.

We see the word badass everywhere nowadays. The word is splashed throughout popular culture in magazines, on T-shirts, or on inspirational posters telling us to "be more badass." But what does it mean? For me, a badass woman stands up for herself, is confident, and is not afraid to challenge the hierarchy, the patriarchy, or conventional thinking. It is these key traits that set the following fictional feminists in this section apart. They are all strong-willed, doing what they believe is right and not worrying about what other people think.

FLEABAG,
FLEABAG

"I sometimes worry I'd be less of a feminist if I had bigger tits."

LESSON LEARNED
Grief is inevitable and there's no right way to deal with it.

PLOT
Fleabag is a modern masterpiece written and starring the wonderful feminist icon that is Phoebe Waller-Bridge. The show is a dark comedy featuring the main protagonist only known as Fleabag, who has experienced a lot of trauma following the death of her best friend. We also learn that Fleabag lost her own mother and is still coming to terms with the void that this has left. The show follows how Fleabag deals with the grief and guilt that come with losing someone you love.

FEMINIST ICON
Fleabag is witty, sexy, and heart-wrenching in equal measures. Centered on this flawed, funny character whose inner monologue is so relatable, the show felt revolutionary when it was first released in 2016. In the very first episode, Fleabag proclaims she is a "bad feminist" after attending a feminist lecture where only she and her uptight sister, Claire, raised their hands when asked the question: "Would you trade five years of your life for the 'perfect' body?"

However, being a bad feminist is simply not true. Fleabag's bond with her sister is proof of this. After the first season, the relationship between the two seems unrepairable when a man comes between them but, just like *Frozen*'s Elsa and Anna, the true love story is the unconditional love both sisters have for one another. Fleabag teaches us that sisterhood may be hard, intense, and messy at times, but it's a bond that runs deep and lasts a lifetime.

Many of the most poignant moments of *Fleabag* come from the rawness of her grief, something those who have lost someone are likely to relate to. Hollywood's portrayal of grief makes it easy to think that if you aren't weeping into a hanky, dressed in a black veil, or making sure everyone else is fine, you are not grieving properly. In truth, everyone grieves differently and it's okay to feel a wide plethora of emotions.

This is what Fleabag illustrates so well, that it's okay to *not* be okay. It's okay to screw up. It's okay to cry and grieve and scream and to be selfish on occasion. It's okay to shut the world out until you're ready to let it back in.

FACTS

- The name Fleabag comes from a real-life nickname that Phoebe Waller-Bridge's family had for her growing up. The real name of the main character is never revealed.
- Fleabag began as a one-woman stage play at the Edinburgh Fringe Festival back in 2013.
- The "hot priest" role was written specifically for Andrew Scott.
- Fleabag had an alternative ending, but Waller-Bridge has remained tight-lipped on what it entails.
- *Fleabag*'s hilarious fourth wall actually serves a deeper purpose for the character, which is realized by the end of season 1. Fleabag, who is deeply suppressing grief from the loss of her mother and best friend, uses these breaks to escape her troubled reality. They become less frequent during season 2 when she has someone she can talk to about her grief.

ANNIE EASTON,
SHRILL

*"Feminism is really just the long, slow realization
that the things you love hate you."*

LESSON LEARNED
Don't let others dictate your self-worth.

PLOT
Shrill is a comedy drama that ran for three fabulous seasons. It follows the story of Annie, a young body-positive woman who wants to change her life but not her dress size. Annie is trying to make it as a journalist while dealing with a society that judges her worth by her waist measurement rather than her merit. The show was written and based on writer and comedian Lindy West's own experiences and book *Shrill: Notes from a Loud Woman*.

FEMINIST ICON
Shrill is a must-watch for anyone looking to absorb some female empowerment. The show revolves around the journey of Annie Easton who has struggled with low self-esteem her whole life. She has been consistently humiliated by people for being a fat woman and has come to expect society to only judge her by her looks. Being fat is something Annie has constantly been told is a bad thing, a failure of hers that she must try to correct. Whether it's her mom suggesting a new diet she should try, her boss calling her cruel nicknames, or her gynecologist inappropriately suggesting gastric bypass surgery, Annie is constantly bombarded with other people's opinions of her body.

As the show progresses, Annie has an epiphany to stop letting others dictate her self-worth. Following an incident in which her abusive boss Gabe calls her lazy and sloppy because of her weight, Annie writes the perfect clap-back article called "Hello, I'm Fat" that goes viral online. Her work is met with support but also backlash from Internet trolls who are enraged and perplexed that a woman could ever be happy being fat! However, this is just the start of Annie's fat acceptance journey which sees her becoming more outgoing, wearing short dresses, and in one scene of pure joy and fun attending a fat-acceptance pool party.

The now infamous "Hello, I'm Fat" article and its editorial response mirror an actual event in Lindy West's career. West began working as a film editor for the Seattle alternative paper *The Stranger* in 2009. Author and activist Dan Savage was also working at the paper as an editor and had given particular attention to the obesity epidemic in America in several of his pieces. In early 2011, West posted an essay entitled "Hello, I'm Fat" on the paper's blog, calling out the callous and unhelpful culture of fat shaming. The blog post was directed at Savage, and Savage posted a lengthy response days later entitled "Hello, I'm Not the Enemy."

This feminist wonder of a show is packed full of inspiration, but the take-home message we can all learn from Annie is that while the journey to self-acceptance and self-love is never easy, it's well worth it!

FACTS

- Annie is played by Aidy Bryant, the brilliant comedian from *Saturday Night Live*.
- The show also covers some other big feminist issues such as abortions, cultural appropriation, and support of LGBTQ+ rights.
- *Shrill* was canceled before the show writers wanted it to be. The final season was disrupted by the global pandemic, which meant the show needed to be wrapped up by the end of season three.

DORALEE RHODES,
9 TO 5

"I'm gonna change you from a rooster to a hen with one shot!"

LESSON LEARNED
We are strongest when we come together.

PLOT
9 to 5 is the perfect feminist comedy which upon release in 1980 felt light-years ahead of its time. The storyline centers around three women: Judy (Jane Fonda), Violet (Lily Tomlin), and Doralee (Dolly Parton) who all work for the same "sexist, egotistical, lying, hypocritical, bigot" boss, Mr. Hart. The three women come together fueled by the harassment, belittlement, and injustices they have experienced to abduct Mr. Hart before using his absence from the office to introduce ideas like job-sharing for working moms, flexible hours, onsite day care, and equal pay. All of which might not sound like a barrel of laughs, but the on-screen chemistry between Jane, Lily, and Dolly, some drug-fueled fantasies about murdering said boss, and some great comedy slapstick make for a Hollywood hit.

FEMINIST ICON
This cinematic classic is an amazing combination of brilliant casting, feminist power, and a dazzling soundtrack! From the very start with the clicking sound of typewriters and one of the most recognizable hits in music we are treated to the iconic opening of a montage of working women of the eighties (prominent shoulder pads, platinum perms, and all). Although the fashion certainly looks dated now, unfortunately the issues at the core of *9 to 5* certainly aren't. With the global gender pay gap expected to take another one hundred years to close, and the explosion of the #MeToo movement, workplace injustice and harassment is still prevalent in today's society.

It is the adorable Texan Doralee Rhodes who is perhaps given the most disturbing storyline when it comes to the misogynistic Mr. Hart. He repeatedly harasses her for sex before spreading a rumor about their nonexistent affair. What makes it more tragic is that many of the other women in the office believe the rumor and Doralee becomes a social outcast. She's ostracized at lunch, shot down when trying to socialize, and her concerns about Hart's constant, casual sexual harassment fall on deaf ears.

Doralee is tarnished in part due to her appearance as a buxom blonde. Her colleagues act upon their own internal misogyny, blaming her for the alleged affair instead of blaming Hart for using his position of power to live out his own sexual fantasies. It is not until Violet is passed over for a promotion that Doralee finds out about the rumor. Rightly outraged, the three women come together and share their own fantasies about how they would get rid of Mr. Hart from the workplace.

The part of Doralee was written specially for Dolly Parton as a smart commentary on Dolly's own persona. Parton is one of the most successful and prolific musicians of all time, as well as a shrewd businesswoman, humanitarian, and saint of all that is rhinestone and good in this world. However, many people in the media industry at the time were more interested in commenting on her physique rather than her brilliance. This theme has been reflected in Dolly's other works such as her song "Dumb Blonde," which includes the lyrics "Just because I'm blonde don't think I'm dumb cause this dumb blonde ain't nobody's fool."

This movie is feminist gold thanks to the fab trio of Jane, Lily, and Dolly, and will continue to be the soundtrack for women everywhere striving for equality in the workplace.

FACTS

- In an effort to promote childhood literacy, Parton founded Dolly Parton's Imagination Library in Sevier County, Tennessee, in 1995. Nowadays, free books are sent to children all over the country as well as to Canada, the UK, and Australia. As of November 2018, the Imagination Library has provided 112,406,659 free books (and counting!) since it was founded.
- The sound of the typewriters in the song "9 to 5" is credited on the track to Dolly Parton's acrylic fingernails.
- Dolly Parton wrote the hits "Jolene" and "I Will Always Love You" on the same day as "9 to 5".
- In 2008, Dolly Parton created a Broadway musical version of *9 to 5* with music of her own creation. It went on to earn four Tony Award nominations.

SOPHIA BURSET,
ORANGE IS THE NEW BLACK

"I'm finally who I'm supposed to be. I can't go back."

LESSON LEARNED
We are who we choose to be.

PLOT
Orange Is the New Black gained high praise in its depiction of the complex women inmates at a fictional federal prison in Litchfield, New York. Based on the real-life memoirs of Piper Kerman, *Orange Is the New Black: My Year in a Women's Prison*, the series focuses on Piper Chapman, a privileged, upper-class, white, bisexual woman who is sentenced to fifteen months in prison for transporting drugs ten years prior. The series goes on to spotlight many of the inmates of Litchfield, often covering difficult and important subjects like violence against women, black lives matter, injustice in the justice system, and queerness.

FEMINIST ICON
There are so many positive things that can be said about the cast of *Orange Is the New Black*. It features multiple characters from minority backgrounds, highlights female friendship, and depicts transgender issues. The latter includes trans inmate Sophia Burset (played by the real-life trans-rights advocate Laverne Cox). Over the first three seasons, Sophia's backstory is revealed. Once a firefighter, Sophia (who's masculine name is Marcus) struggles internally with keeping up the gendered roles expected of her. The pressure Sophia is under in hiding her true self is something most trans people experience. Unfortunately, Sophia's struggles don't stop there. While in Litchfield prison, Sophia's hormone therapy is stopped due to budget cuts and she is repeatedly denied the right to see a doctor by the prison warden, as he did not consider her health needs as an emergency. Sophia is forced to take matters into her own hands, self-harming in order to see a medical professional. This is unfortunately the case for many incarcerated women, including famously Chelsea Manning.

Sophia is also a savvy businesswoman, running the prison hair salon, where she is a sympathetic ear to many of the prisoners. She allows the women to feel like themselves again through self-care and empowering advice. It is in this salon setting where we see Sophia's sense of humor really pop!

Ultimately, the pain that haunts Sophia the most is the lack of acceptance she experiences from her own family. She eventually finds peace within herself once her son, Michael, comes and visits her in prison. If there is a moral to Sophia's storyline, it's that a woman, no matter her background, should not have to prove to anyone else that she is a woman. Her own voice should be proof enough.

FACTS

- The series was written and created by Jenji Kohan who is no stranger to creating wonderfully in-depth female-centric storylines, being a writer and producer on shows like *Gilmore Girls* and *GLOW*.
- Sophia's storyline in *Orange Is the New Black* is based on real-life trans inmate CeCe McDonald, who spent nineteen months in a men's prison for defending herself during a racist, transphobic attack.
- Laverne Cox's portrayal of Sophia Burset made history, as she became the first trans individual to receive an Emmy nomination for her role in the series.
- The first season of *Orange Is the New Black* claimed twelve Emmy nominations in 2014 and won three, making it one of the first critically acclaimed series for streaming giant Netflix.

PART NINE
Heroines of Herstory

The majority of history has been written by white men, about white men, for other white men. Because women have not been reflected in the annals of time, it is hard to image how 50 percent of the population experienced an era, how they dealt with changing times, and the many contributions they made along the way. It's often said that silence is golden, but for women it has equaled eradication from the history books.

These limited resources about women in history have trickled down into the media and pop culture. A 2018 survey[*] covering twelve language versions of Wikipedia found that 90 percent of contributors reported their gender as male, while only 8.8 percent identified as female and 1 percent as other. In addition, a 2021 study[†] found that 41 percent of biographies nominated for deletion were women despite only 17 percent of published biographies being about women.

Therefore, this section is all about shining the spotlight on feminist narratives throughout history. From Agent Carter setting the world right in post-war America to Tracy Turnblad living her best life in the swinging sixties, all the fictional feminists in this section are deemed to make a mark on history.

---

[*] Community Engagement Insights 2018: Fostering learning to improve support to Wikimedia communitie. (Accessed March 2022)
[†] Tripodi, F., Ms. Categorized: Gender, notability, and inequality on Wikipedia. *New Media & Society*. June 2021.

PEGGY CARTER,
AGENT CARTER

"I know my value."

LESSON LEARNED

Don't let a past relationship define you.

PLOT

Agent Peggy Carter first made her appearance in the Marvel Universe as the love interest of Captain America, but thankfully she is so much more than a sexy superhero's side piece. Set in New York in 1946, Peggy works for the Scientific Strategic Reserve (SSR), an organization so strife with sexism it makes the TV series *Mad Men* look almost woke. The series follows Peggy having to balance the routine office work she does while secretly assisting Howard Stark (Iron Man's father) who has unjustly become an enemy of the United States.

FEMINIST ICON

Forget James Bond. If you need a secret agent, Agent Carter's your gal! She is tough, intelligent, and compassionate, yet also impulsive and stubborn, making her realistic rather than a flawless ideal of what a female superhero should be. Carter is often overlooked by her colleagues, perceived as being weaker than the other male agents. She is often expected to do the "womanlier" tasks in the office like filing and making coffees for everyone. Although the show is set more than seventy years ago, a lot of the issues Peggy has to deal with are still unfortunately relevant in the workplace today. But just like Peggy, we know our own value and will press on despite the patriarchy's best efforts!

At times, the validity of Carter's position is questioned due to her past relationship with Captain America. However, Peggy has the confidence not to let other people's opinions hold her back. She knows she is not defined by a man.

Peggy is also a fantastic ally. Throughout the show she defends the rights of other women and supports them, proving that there is more to being a superhero than magical or alien powers.

FACTS

- Hayley Atwell who portrays Agent Carter uses her platform to advocate for women's rights. When asked about sexism within the series, Hayley replied, "We've come a long way. We've come a lot further than Peggy in 1946, but we've still got a long way to go. In terms of equality in the workplace, equal pay for women, we're still struggling in the Western world. We're still struggling for women, not to mention globally. With the tremendous insight we've had over the past few years, the actual brutal suffering of women all over the world, there's an actual need for women's education, and women's rights."

- The show gained much acclaim for its amazing costumes and set designs. Costume designer Gigi Ottobre-Melton was keen to make each costume historically accurate as well as aesthetically pleasing, ensuring that Agent Carter always looked the definition of glamor and class.
- Agent Carter was first introduced to the Marvel Universe in 1966 in a Captain America comic called Tales of Suspense #75.
- In the Marvel Universe, Peggy Carter became the director of S.H.I.E.L.D. in 1970.

TRACY TURNBLAD,
HAIRSPRAY

"People who are different, their time is coming!"

LESSON LEARNED
Follow your dreams and have fun along the way.

PLOT
The musical *Hairspray* is set in 1962 in Baltimore and follows the gorgeously curvy teen Tracy Turnblad. Tracy has one dream in life and that is to dance on *The Corny Collins Show*. She auditions for a spot and wins, becoming an overnight celebrity. She hopes to use her influence to topple Corny's reigning dance queen and bring racial integration and body positivity to the show.

FEMINIST ICON
While some musicals have outdated gender roles or homophobic content, *Hairspray* rights this wrong by putting its female characters at the heart of the story, giving hope to feminism on Broadway! Tracy Turnblad is the inspirational character we all need in our lives. She is passionate and optimistic, believing there are endless possibilities for great things to happen every day. She remains positive even when mean girl Amber and her bigoted mother Velma try to stop Tracy from joining *The Corny Collins Show* because she isn't the stereotypical skinny girl the show normally casts. Tracy radiates the body positivity that was way ahead of her time. As the glamorous Motormouth Maybelle would say, "Who wants a twig when you can climb the whole tree?"

After earning her spot on *The Corny Collins Show*, Tracy proves that being different is a good thing and that times are changing for the better. She pushes for integration on the show after meeting her friend Seaweed's younger sister Little Inez who is an incredibly talented dancer and singer. Tracy joins a protest against segregation to allow black dancers on the show, even though she knows she could lose her spot by doing so.

Tracy doesn't just walk through life, she dances through it, with an effortless ease that I've always admired. In other cult teenager dramas, the female lead usually starts off as a shy awkward girl who gets a makeover and is suddenly the talk of the school. This is not the case for Tracy. She does not need a magical transformation to shine or get the guy of her dreams. For once, the guy undergoes the magical transformation when

Tracy's crush wakes up to what an amazingly strong woman she is! Tracy is an inspiration for everyone living in a constantly evolving world.

FACTS

- John Waters's original 1988 version of *Hairspray* starring Ricki Lake and Divine was groundbreaking in many ways. This landmark film is one of few to champion body positivity in Hollywood with its strong lead protagonist at a time when this issue was not widely covered. This is the first of Waters's films to become widely popular outside of gay and counterculture circles.
- The 2007 film starring Nikki Blonsky, Zac Efron, and John Travolta is one of the highest grossing musicals of all time.
- Michelle Pfeiffer and Christopher Walken were in the film at the request of John Travolta, who insisted they join the cast.
- The original Broadway production used over 150 wigs! However, Tracy only wears three throughout the show.
- Following in the tradition of John Waters's original film, both the play and the 2007 movie showcase a number of costumes from vintage outfits that designers bought from secondhand shops, antique stores, and flea markets. This was done to evoke a more authentic feel of the era.

ORLANDO,

ORLANDO: A BIOGRAPHY

"As long as she thinks of a man, nobody objects to a woman thinking."

LESSON LEARNED

Gender is no longer binary, but part of a fluid continuum.

PLOT

Orlando: A Biography is the story of an eponymous hero who is born as a male nobleman in England during the reign of Queen Elizabeth I. He undergoes a mysterious change of sex at the age of thirty and lives on for more than *three hundred* years to the birth of Queen Elizabeth II. Orlando has many adventures during their lifetime, from forming a dalliance with Elizabeth (the first, not the second), pursuing an ill-fated romance (while still a man) with a Russian princess, serving as ambassador in Constantinople, and (now as a woman) living with gypsies before returning to England.

FEMINIST ICON

Virginia Woolf's *Orlando* touched on key aspects of sexual identity that would have seemed totally alien at the time of publication. Woolf's forward-thinking vision along with some of the most masterful writing I have ever read in *Orlando* shows her true genius as well as how ahead of her time she really was. For these reasons, the character of Orlando has become an iconic LGBTQ+ character in fiction. On seeing himself as a herself for the first time in the mirror, she remarks, "Different sex. Same person." What sounds a simple statement of fact is one so many still struggle to comprehend today.

One of my favorite scenes in *Orlando* comes in the final chapter when Orlando reminisces about all the different people they have been over the years. From a lover to an explorer, a wife to a poet, Orlando has managed to be everyone, but more importantly, in this moment they accept their own faults as much as their assets. To embrace every aspect of yourself is a hard lesson to learn but one we should all aspire to do.

She may have been married to Leonard Woolf, but it is no secret that Virginia Woolf was romantically involved with women, too. In particular, poet Vita Sackville-West, who is said to have been the inspiration behind the Russian princess in *Orlando*. It is an affair so well-known and documented that the letters between the two writers are a favorite topic for academics to this day. Woolf's exploration of gender and sexuality both in her real life and in her writings helped shape the ways in which feminism and female sexuality has been defined in the literary world.

ABOUT THE AUTHOR

It's undeniable that Virginia Woolf is one of the most groundbreaking feminists of the twentieth century. In Woolf's collection of essays, *A Room of One's Own*, she asserts the importance of a woman's ability to be financially independent for her own emancipation, something which was not easy to achieve at the time.

FACTS

- The final sentence in *Orlando*, "the twelfth stroke of midnight, Thursday, the eleventh of October, Nineteen hundred and Twenty Eight" was the exact date and time of the book's publication.
- Woolf's first published piece of writing was about the Brontë sisters, which appeared in the *Guardian* in 1904. She idolized the literary Brontë sisters and wrote the piece after a pilgrimage to their family home, that almost holy shrine where the classics *Jane Eyre*, *Wuthering Heights*, and *The Tenant of Wildfell Hall* were all written.
- Woolf struggled with mental health issues her whole life. It is now suspected that she had anorexia, bipolar disorder, and severe depression. However, as these diseases were not understood at the time, Woolf tragically took her own life in 1941.
- Woolf was a member of the Bloomsbury Group, a collection of associated English writers, intellectuals, philosophers, and artists who were extremely forward thinkers of the time. They rejected Victorian ideals and supported gay rights, women in the arts, pacifism, and free love.

JO MARCH,
LITTLE WOMEN

*"I'm happy as I am, and love my liberty too well to
be in a hurry to give it up for any mortal man."*

LESSON LEARNED
Treasure your talents. They may prove useful for yourself and those you love.

PLOT
The story follows the lives of the four March sisters: Meg, Jo, Beth, and Amy. The four of them endure the harsh New England winters with their mother while their father fights in the Civil War. Together, they cope with growing up, growing apart, falling in love, and chasing their dreams.

FEMINIST ICON
Sisters Josephine, Meg, Amy, and Beth were as different as sisters could be. Meg is the beautiful one, both inside and out, Beth is the peacekeeper of the sisters, holding everyone together, while Amy is everyone's spoiled younger sister. I always wanted to be the tomboyish, headstrong heroine Josephine, or Jo as she preferred to be called.

Jo March was a brilliantly bold and bright young woman who refused to conform to the binding social constraints of Civil War–era New England. While many of the other women in this book complied with the period's conventional ideals of being mild, compliant, demure, and feminine, Jo wasn't having any of it. Eventually, she was rewarded for her intrepid behavior and achieves her dream of gaining literary success. Jo showed me that it's okay to pursue challenging dreams.

ABOUT THE AUTHOR
Louisa May Alcott, born in New England in 1888, was a novelist, short-story writer, and poet. She had an exceptional upbringing and was always surrounded by well-known writers and thinkers of the time, including Ralph Waldo Emerson, Nathaniel Hawthorne, Henry David Thoreau, and feminist journalist Margaret Fuller; all of whom were family friends who perhaps influenced Alcott. If there was any doubt as to whether Alcott was a feminist, she was the first woman to register to vote in Concord, Massachusetts, in 1879. She remained unmarried throughout her life and valued her independence above all else.

FACTS

- *Little Women* took Alcott just ten weeks to write. She became consumed by the novel and worked on it day and night.
- *Little Women* was originally published in a two-part installment.
- Alcott refused to have Jo marry Laurie. She noted in her journal, "Girls write to ask who the little women will marry, as if that was the only end and aim of a woman's life."
- Alcott wrote two sequels to *Little Women*: *Little Men* in 1871, and *Jo's Boys* in 1886.

PART TEN
Wonderful Warrior Women

Throughout human history, women have defied stereotypes and become the protectors, leaders, and fighters of their communities. From Boudica to Joan of Arc, powerful stories of real-life warrior women have gone on to influence folklore and mythology with fantastic tales of women holding more "masculine" roles, emphasizing women's agency and capacity for power.

LARA CROFT,
TOMB RAIDER FRANCHISE

"A scar means I survived."

LESSON LEARNED

Never apologize for being a powerful woman.

PLOT

Lara Croft is one of the most successful video-game heroines of all time. Lara is a very well-to-do British aristocrat who follows in her father's footsteps as an archaeologist, treasure hunter, and tomb raider. She plunders other countries' treasures and returns them to her own manor in order to keep powerful objects away from other people who might be hunting them.

FEMINIST ICON

Since appearing on the scene in 1996 with the first installment of *Tomb Raider* for Sony's PlayStation, Lara has been a cultural flash point in a very male-dominated gaming industry. Originally branded as a sex symbol, her grossly disproportional body with huge polygonal breasts received a great deal of criticism. Only a male game designer could give a woman breasts that large and expect her to not only run but high kick and jump her way through abandoned jungle caves. Fortunately, Lara has had her own twenty-first-century feminist reboot and the focus is now on her intelligence and athleticism rather than her body proportions.

When Paramount Pictures announced that they would be making a *Tomb Raider* movie, it was Angelina Jolie who brought the role to life first and insisted on doing almost all of her own stunts. In the 2001 movie, I was obsessed with the amazing bungee dance scene, where Lara masterfully defeats some home invaders while all the time soaring around her massive mansion like a badass aerial ballerina.

FACTS

- Lara Croft has sold over 58 million video game units worldwide.
- Lara Croft holds six Guinness World Records, including Most Successful Video Game Heroine and Most Recognizable Female Character in a Video Game.

- Lara Croft is not only a talented archeologist, but she also had a brief career as a pop star, with two albums under her belt. However, one album entitled *Female Icon* included ill-considered singles such as "Getting Naked," "Sure is Pure," and "Feel Myself." Recorded in English, producers soon woke up to what a horrific mistake they had made, and the albums were buried, only ever released in France.
- Lara Croft is based on a real woman who lived in Derby, England, at the time the game was being written. The developers liked the name Lara for the main protagonist but were undecided on a last name that sounded British enough. The decision was made to look through a phone book and they stumbled upon the real-life Lara Croft.

OKOYE,
BLACK PANTHER

"Sisters, you are here to learn to serve, to fight, to be fierce, to be fearless."

LESSON LEARNED
Trust and loyalty should always be earned, never given.

PLOT
Black Panther was a mega Hollywood hit that felt many years overdue in the white male-dominated genre of superhero movies. It is a fairly classic superhero origin story, with a death of a king from a faraway land forcing a young male protagonist to take on responsibility of a kingdom and a new superhero alter-ego, all the while defeating foes and securing the safety of his people. However, *Black Panther* is anything but the overplayed tropes that we have come to expect of this genre, with some amazing casting and stand-out strong female roles such Queen Ramonda, Shuri, Okoye, and Nakia.

FEMINIST ICON
Black Panther broke the Bechdel test and has one of the largest casts of women. Key female characters are Queen Ramonda, Shuri, Okoye, and Nakia. These four hold many conversations among themselves that do not revolve around male protagonist T'Challa (although a lot do).

However, Okoye the most fearsome warrior in the whole of Wakanda and head of the Dora Milaje, the all-female Wakandan special forces that serves as an elite bodyguard to the royal family, has to be up there. Although her skill as a fighter is second to none, it is Okoye's loyalty to her king and country that earns her the rank of general. One of her most testing moments is when she confronts her longtime love W'Kabi on the battlefield, who questions her willingness to strike him down. "For Wakanda?" she answers in defiance. "Without question." Her driving force is not a man, but a nation she loves and is proud of. This is why she chooses to fight.

In spite of having no superpowers, Okoye is one of the most powerful characters in the Marvel franchise. She wields her womanhood as a strength, casting femininity in a whole new light in the process. It is not one or the other with Okoye. It can be both. Though her bald, tattooed head is rooted in her Wakandan traditionalism, Okoye still shatters our real-world standards of traditional beauty through a fierce kind of fabulousness. In one scene from the film while Okoye is undercover on a mission in South Korea, she wears a wig. However, she later throws her wig off and uses it to literally take down a white man, showing that she is in charge.

Okoye is a trailblazer for women and girls, destroying stereotypes and leading the charge for black women moving from the margins to the center of the narrative. Here's hoping to see a lot more of these amazing women. Wakanda forever!

FACTS

- Actress Danai Gurira who plays Okoye agreed to play the part without having even read the script. She was sold on the concept alone.
- Gurira also commented that part of playing a character in the Dora Milaje meant that she and fellow actresses had to completely shave their heads, an act she found empowering, helping her bond with her castmates in the process.
- *Black Panther* was the first Marvel movie to win Academy Awards.
- The Academy Award for best costume design went to Ruth E. Carter (the first black woman to win in this category), who created more than one thousand costumes for the film. She drew inspiration for the costumes from Afrofuturism, Afropunk fashion, and traditional African tribal garments. Her costume design for the warrior women of the Dora Milaje gained particular praise as being practical and protective and not just a glorified bikini that Hollywood often deems appropriate for female warriors.
- Visibility and inclusivity were important both on and off the screen. Director Ryan Coogler made sure a lot of the key department heads off-screen were African American, women, or both.
- Adoption of black cats reportedly increased after the film's release.

DIANA PRINCE,
WONDER WOMAN

"If no one else will defend the world, then I must."

LESSON LEARNED
There is strength in the truth.

PLOT
Published by DC comics, Princess Diana of Themyscira (also known as Diana Prince) is a fictional superhero and one of the founding members of the Justice League. From a secluded nation of Themyscira, she is a descendant of royalty and gods and grew up with all female warriors championing peace and fighting for justice and equality in a man's world.

FEMINIST ICON
You can't have a list of fictional feminist icons, let alone warrior women, and not mention the Amazonian warrior princess that is Wonder Woman. Few other fictional characters have had such a cultural impact on the world. First exploding onto the scene in the early forties with her debut in All-Star Comics #8, she was an instant trailblazer. At a time when women in comics were meek or nonexistent, Wonder Woman was different. She was a butt-kicking, take-charge champion of justice who could hold her own with the likes of Batman and Superman. Although the road to feminism for Wonder Woman has had a few bumps (we will get to those later), it says a lot that she is still enchanting audiences eighty years on.

Created by psychologist William Moulton Marston, a man who lived in a polyamorous relationship with two feminists and was inspired by the suffrage movement, Diana of Themyscira has many an attribute to celebrate. While I could wax lyrical about her ability to fight or to command, one of her most valuable assets that defines her as a superhero is her compassion. In both the comics and in the films, Diana routinely experiences moments of compassion and empathy which would floor her fellow Justice League colleagues like Batman (who has the emotional range of a teaspoon). It is no coincidence that her weapon of choice is a magical lasso that compels those ensnared to tell the truth. Because Diana was raised and trained with a duty to protect others, her motives are purer than others in the comic book world.

Like many female figures, Wonder Woman's image has been oversexualized in the past, but the evolution of her character reflects our own evolution of how we view gender, femininity, war, and inclusivity today.

FACTS

- The "Wonder Woman pose" a so-called power pose that involves standing with your legs shoulder-width apart, chest out, arms placed on your waist, has been scientifically proven by psychologists to improve confidence and self-esteem. This trick has reportedly been used by public figures and political leaders across the globe.
- Patty Jenkins directed the most recent Wonder Woman films, which not only ace the Bechdel test but also received much praise in casting Gal Gadot to pick up the golden tiara. Gadot was pregnant while filming the first film, which was seen as an important step in cracking the glass ceiling and removing obstacles in the way of women in the film industry.
- Patty Jenkins and costume designer Lindy Hemming made sure the warrior women of Amazonia had adequate armor, with metal chest plates and navels covered (something you would want if you were going into battle).
- In 2016, the United Nations made a curious appointment of making Wonder Woman the global organization's Honorary Ambassador for the Empowerment of Women.
- Wonder Woman has her own day, October 21.

MULAN,
DISNEY'S *MULAN*

"My duty is to my heart."

LESSON LEARNED
There is honor in fighting for what you believe in.

PLOT
The 1998 film *Mulan* sees the Huns along with their leader, Shan Yu, invading China. Out of fear for her injured elderly father going to war, Mulan disguises herself as a man (Ping) and takes his place. Mulan's ancestors send her an adorable companion, a dragon named Mushu to keep her company. Her commanding officer Li Shang initially asks her to leave, but through her perseverance, Mulan learns to fight, befriends her fellow soldiers, and destroys the Hun army (and some gender stereotypes along the way!).

FEMINIST ICON
Mulan's first appearance was in the poem "The Ballad of Mulan" some 1,500 years ago, written during the Northern Wei Dynasty when ancient China was ruled by non-Han Chinese ethnic groups and divided between the north and south. The poem tells the story of a woman who disguises herself as a man to take the place of her father when he's drafted into the army. The story of Mulan has been retold countless times, and throughout the history of China, Mulan has been manipulated into becoming a mouthpiece to enforce the idea of nationalism and obedience to one's ancestors.

However, like many, my first experience of the story of Mulan was through the 1998 smash hit Disney animation. Unlike most other Disney princesses, Mulan was an awkward, self-doubting heroine that many people (myself included) could actually identify with! She defied gender roles, destroying the notion of a damsel in distress, and challenged the patriarchy every step of the way.

One of the most iconic gender deconstructing scenes in Mulan is the transformation of Mulan from a woman to a man (Ping). We see Mulan cutting her hair and dressing as a man, leaving her flower comb (a feminine object) in place of her father's order to serve (a masculine object). Mulan's strength not only comes with learning to fight and defending family honor, but her strength in influencing those around her for the better, whether it is her fellow soldiers, her army general, or the Emperor of China himself. She is a warrior, yes, but she is a leader and a trailblazer foremost.

FACTS

- It is not clear whether the legendary warrior Hua Mulan is indeed a real person. However, her tale has fascinated generations with many plays, poems, and works of art dedicated to her tale.
- In its earliest stages, *Mulan* was going to be a straight-to-video animated short called "China Doll" about an oppressed young woman in China who finds happiness after a British soldier sweeps her off her feet (and out of China). None of Disney's top animators wanted to work on it, thankfully, and the idea eventually turned into the tale of Mulan, the independent heroine.
- Disney animators often pull characteristics from the voice actors when designing their characters, and Ming-Na Wen (who played Mulan) was no exception. After noticing Wen had a habit of touching her hair, the artists decided that Mulan would, too.
- Mulan is often credited with launching Christina Aguilera's career. The former Mickey Mouse Club member released a version of "Reflection" in 1998, which would become her first-ever chart success.

XENA,

XENA: WARRIOR PRINCESS

"A strong person is one who is quiet and sheds tears for a moment, and then picks up her sword and fights again."

LESSON LEARNED

True love is finding your soulmate in your best friend.

PLOT

As stated in the theme song of *Xena: Warrior Princess*, "In a time of ancient gods, warlords, and Kings, a land in turmoil cried out for a hero." Xena is that hero who is on a quest to redeem herself from a dark past by using her formidable fighting skills for good and to help others. The series is set in ancient Greece with many of the characters and enemies encountered coming from Greek mythology.

FEMINIST ICON

This nineties TV hit is packed full of strong female characters and is well overdue for a rewatch (or if any Hollywood execs are reading, a reboot!). The legacy of *Xena: Warrior Princess* has grown to become a feminist and lesbian icon. Xena, who is every bit the badass you would expect, uses her own wit and strength to save the day.

The show not only smashes the Bechdel test but also features one of the most progressive female friendships in the action genre to date between Xena and Gabrielle. The pair, who start off as strangers, travel the land helping others and also developing and growing as characters themselves. Xena and Gabrielle fall out and fight with each other like all close friends do but they always come back stronger. Xena's friendship circle includes many strong women from throughout history who all make an appearance, such as Cleopatra and Boudicca. She is not intimidated by other successful women or worried about others stealing her spotlight. Xena is a true feminist, knowing that success is when everyone can succeed, not just the individual.

There has been much speculation about the nature of the relationship between Xena and Gabrielle and it is something which is never directly resolved during the show. Many saw the on-screen bond of Xena and Gabrielle as more than just friendship, later confirmed by actress Lucy Lawless who played Xena in the show.

FACTS

- The show went on to become much more popular than its predecessor *Hercules*, and Hercules star Kevin Sorbo was not pleased about it. In the book *Hercules: An Insider's Guide to the Continuing Adventures*, the actor complained about Xena becoming Hercules's "physical equal."
- Despite the empowering storytelling going on in front of the camera, only five of Xena's 134 episodes were directed by women.
- When Lawless got pregnant with her second child during the show's fifth season, the producers accommodated their star's major life event by writing it into the show. And that's how Eve, Xena's daughter, came into existence.
- Xena's weapon of choice—her chakram—is a real weapon that dates back to second century BCE India.

PART ELEVEN
Galactic Gals

Space may be the final frontier, but unfortunately for women it has so far been just as male dominated up there as it is down here on Earth! However, there are some stand-out heroines who are making their mark on sci-fi audiences throughout the galaxy. For example, take the popular British sci-fi series *Doctor Who*. After fifty-seven years and twelve previous regenerations of the Doctor, 2019 saw the casting of the first female Doctor. The decision was met with what seemed like equal parts joy and equal parts hate. Joy from many that finally a show that features unimaginable monsters and aliens from across the galaxy was also able to image a female lead who was not just a sidekick love interest or a tragic plot twist to be killed off and replaced at the end of a season. But just like the hate-filled Daleks, there were those who saw a female doctor as an encroachment on all they hold dear in the world.

ELLEN RIPLEY,
ALIEN FRANCHISE

"Get away from her, you bitch!"

LESSON LEARNED
Motherhood comes in many shapes and sizes.

PLOT
Back in the first movie of the franchise (*Alien*), Ripley is introduced as a warrant officer aboard the *Nostromo*, a spaceship en route to Earth. Having been placed in stasis for the long journey back to Earth, the crew is awakened when the ship receives a transmission from an unknown origin from a nearby planet. Following the distress signal and landing, an unknown creature (Xenomorph) infiltrates the ship, slowly killing every crew member until Ripley is the last woman standing. Ripley escapes the doomed ship on a shuttle, only to discover the creature is also aboard. However, Ripley manages to expel the Alien into space and puts herself back into stasis to return home. Unfortunately for Ripley, this is not her only encounter with these creatures. Throughout the *Alien* franchise, she's stripped of all military ranking, almost raped, watches Xenomorphs kill everyone she knows, gets impregnated with a Chestburster, and throws herself into a furnace to prevent its birth, only for scientists to bring her back to life two hundred years after her death. It's a lot for one woman to go through!

FEMINIST ICON
It has since been revealed that screenwriter Dan O'Bannon first wrote the character of Ripley as a man. However, when *Alien* came out in 1979, it was a year of male-dominated action movies, including *Apocalypse Now*, *Rocky III*, and *Moonraker*. *Alien* needed a gimmick, something to stand out from the crowds, so Ripley the kick-ass alien-killing heroine was born.

There are many things that make Ripley a strong female lead. She is a survivor in every sense of the word and, armed with a flamethrower, she is unstoppable! But in contrast to most horror/action films, what makes Ripley truly strong is her compassion. She unashamedly displays her instincts to nurture and protect Newt, which becomes her driving force and source of strength throughout the film.

When the aliens take Newt, all Ripley can think about is getting her back. It doesn't matter if she has to put her own life in jeopardy. She heads straight into the aliens' nest, back into her own nightmares, without even a second's hesitation. Ripley's compassion doesn't make her a weak or vulnerable character. It's her driving force.

FACTS

- All the aliens in the movie were designed by surrealist painter H. R. Giger. O'Bannon handpicked him for *Alien*.
- The studio originally requested that Ellen Ripley was written out of the sequel due to a pay dispute with the actress who plays her, Sigourney Weaver. Instead, thanks to the director's persistence, they ended up paying Weaver thirty times more than what she earned for *Alien*.
- Spacesuits (even fake ones) tend to get hot—especially when they don't let any air out. Add in set lighting and a summertime production schedule and you have some truly sweltering conditions. Veronica Cartwright, who played Lambert, revealed in *The Beast Within: The Making of Alien* that the actors were fainting so regularly that a nurse was kept on standby with oxygen tanks.

PRINCESS LEIA ORGANA, STAR WARS FRANCHISE

"Well, I guess you don't know everything about women yet."

LESSON LEARNED

Power is not given to you. You have to take it.

PLOT

George Lucas's Star Wars Universe covers twelve films (with more in various stages of production), at least half a dozen television series, dozens of video games, hundreds of books and comic books, and enough merchandise to sink the Death Star. The timeline of all these stories is rigid, vast, and at times daunting. The fourth, fifth, and sixth movies take place, chronologically, before the first, second, and third. A bold move for a relatively new filmmaker, but George Lucas wanted the audience to dive straight into the center of his epic space opera. This decision not to have a singular starting point has allowed the Star Wars franchise to become ever-expanding, growing far beyond the movies Lucas had originally planned.

Strong female protagonists are being featured more and more in the Star Wars Universe, with Rey, Jyn Erso, Ahsoka Tano, and Cara Dune headlining in recent film and TV series. It looks like the battle with the patriarchy has been won in space—only the Empire left!

FEMINIST ICON

Intelligent, powerful, witty and gets the job done, Princess Leia is the epitome of a feminist icon! Much of Princess Leia's fight and defiance must be attributed to the amazing feminist behind the character, Carrie Fisher. Carrie, who was only nineteen years old when she first appeared in *A New Hope*, was never afraid to call out the relentless double standards she witnessed in Hollywood throughout her career. Her witty put-downs and no-nonsense attitude make Carrie just as much a force to be reckoned with as Princess Leia herself.

Leia never lets anyone intimidate her and she stands up for what she believes in. She treats those around her with great respect, acknowledging their effort and not being afraid to show emotion. She does not simply act like a man in charge, she keeps her femininity and rocks some iconic looks along the way.

Leia does not represent the fairy-tale ending we have come to expect from other princess stories we grew up with. Yes, she was with Han Solo, but when she sees the relationship isn't working anymore, she leaves him (even though she continues to love him). It is these bittersweet moments that make Leia such a real and relatable character.

FACTS

- Even though Leia is the daughter of Darth Vader, one of the most powerful users of the Force, we don't really get to see her full abilities to wield the Force in the films. In recent Marvel comics, however, Leia is shown using the Force to look into the past, and elsewhere she can wield a lightsaber.
- Leia's famous white dress didn't allow for a bra underneath, and Fisher's breasts had to be held down with gaffer tape. "As we all know," Fisher later joked, "there is no underwear in space."
- In Carrie's 2016 autobiography *The Princess Diarist*, Carrie had the following to say about sexism in Hollywood: "The crew was mostly men. That's how it was and that's pretty much how it still is. It's a man's world and show business is a man's meal, with women generously sprinkled through it like overqualified spice."
- Leia's "bagel bun" hairstyle from *A New Hope* is still one of the most iconic hairstyles in movie history. The look was inspired by how women used to set their hair in turn-of-the-century Mexico.

LIEUTENANT NYOTA UHURA, STAR TREK FRANCHISE

Sulu: *"I'll protect you, fair maiden."*
Uhura: *"Sorry, neither."*

LESSON LEARNED

Be the change you want to see.

PLOT

Star Trek set in the twenty-third century followed the crew of the starship *Enterprise*, a multicultural, multispecies group of individuals representing the United Federation of Planets on a mission to explore strange new worlds and civilizations. Lieutenant Nyota is a translator and communications officer on the starship *Enterprise* who specializes in linguistics, cryptography, and philology.

FEMINIST ICON

Lieutenant Nyota Uhura was the first black female character to be portrayed as a figure of authority on an American television series. She was not in a subservient position but was strong, confident, capable, and a complex human being.

Uhura was played by actress Nichelle Nichols in the original series. Nichols endured a considerable amount of racist harassment for her role as a leading cast member on *Star Trek*. When she wanted to quit the show because of this, it was Dr. Martin Luther King Jr., a massive fan of *Star Trek*, who implored Nichols to remain, saying "You can't [quit]. You're part of history." He told her that she was playing a vital role model for children and young women across the country who would see black people being treated as equals. Nichols was convinced and continued playing the role for twenty-five years.

Seeing Nichelle Nichols playing Lieutenant Uhura did inspire many young girls, not only to be actresses but also scientists and astronauts. Mae Jemison (the first black woman in space) would begin her shifts by communicating to Mission Control in Houston that "hailing frequencies were open" (Uhura's famous line). This one example beautifully illustrates the influence that strong fictional women can have over our lives.

Actress Nichelle Nichols leveraged her character's popularity to influence the real-life space program. After a 1975 NASA convention showed the homogeneity of NASA's white, male-dominated astronaut pool, Nichols decided to take action to trigger the change she was meant to represent and wrote to newspapers and magazines highlighting why the pool of American astronauts does not represent the people of America.

FACTS

- Nyota Uhura means "star freedom" in Swahili.
- The first American interracial kiss was shown on *Star Trek* between Uhura and Captain Kirk (William Shatner) in 1968. During this period in history, there was still a massive taboo against interracial relationships. Fearing that the kiss may "upset" Southern viewers, National Broadcasting Company (NBC) executives wanted to capture an alternative scene without the kiss. However, Shatner sabotaged the scene without the kiss by crossing his eyes directly into the camera, making the take unusable and ensuring the kiss aired.
- One of Uhura's best-known quotes on the show was "Hailing frequencies open, sir." After seeing the phrase so often repeated in the script, Nichols said, "If I have to open hailing frequencies one more time, I'll smash this goddamn console!"

Index

NOTE: Fictional characters are alphabetized by first name.

ALSO AVAILABLE

The Science of Aliens: The Real Science Behind the Gods and Monsters from Space and Time | *by Mark Brake*

The Science of the Big Bang Theory: What America's Favorite Sitcom Can Teach You about Physics, Flags, and the Idiosyncrasies of Scientists | *by Mark Brake*

The Science of Doctor Who: The Scientific Facts Behind the Time Warps and Space Travels of the Doctor | *by Mark Brake*

The Science of Fortnite: The Real Science Behind the Weapons, Gadgets, Mechanics, and More! | *by James Daley*

The Science of Harry Potter: The Spellbinding Science Behind the Magic, Gadgets, Potions, and More! | *by Mark Brake & Jon Chase*

The Science of James Bond: The Super-Villains, Tech, and Spy-Craft Behind the Film and Fiction | *by Mark Brake*

The Science of Jurassic World: The Dinosaur Facts Behind the Films | *by Mark Brake & Jon Chase*

The Science of Minecraft: The Real Science Behind the Crafting, Mining, Biomes, and More! | *by James Daley*

The Science of Monsters: The Truth about Zombies, Witches, Werewolves, Vampires, and Other Legendary Creatures | *by Meg Hafdahl & Kelly Florence*

The Science of Science Fiction: The Influence of Film and Fiction on the Science and Culture of Our Times | *by Mark Brake*

The Science of Serial Killers: The Truth Behind Ted Bundy, Lizzie Borden, Jack the Ripper, and Other Notorious Murderers of Cinematic Legend | *by Meg Hafdahl & Kelly Florence*

The Science of Star Trek: The Scientific Facts Behind the Voyages in Space and Time | *by Mark Brake*

The Science of Star Wars: The Scientific Facts Behind the Force, Space Travel, and More! | *by Mark Brake & Jon Chase*

The Science of Stephen King: The Truth Behind Pennywise, Jack Torrance, Carrie, Cujo, and More Iconic Characters from the Master of Horror | *by Meg Hafdahl & Kelly Florence*

The Science of Strong Women: The True Stories Behind Your Favorite Fictional Feminists | *by Rhiannon Lee*

The Science of Superheroes: The Secrets Behind Speed, Strength, Flight, Evolution, and More | *by Mark Brake*

The Science of Time Travel: The Secrets Behind Time Machines, Time Loops, Alternate Realities, and More! | *by Elizabeth Howell, PhD*

The Science of Witchcraft: The Truth Behind Sabrina, Maleficent, Glinda, and More of Your Favorite Fictional Witches | *by Meg Hafdahl & Kelly Florence*

The Science of Women in Horror: The Special Effects, Stunts, and True Stories Behind Your Favorite Fright Films | *by Meg Hafdahl & Kelly Florence*